Changing the Food Game lays down the problem and the solution. The book shows clearly why systemic leadership, innovation and collaboration are the only way forward in dealing with the global food problem.

Herman Wijffels, economist and former Dutch representative at the World Bank; Professor of Sustainability and Societal Change, University of Utrecht, The Netherlands

Sustainable food systems are pre-competitive. We must work together to produce more with less by focusing globally on productivity and efficiency while reducing waste. This book suggests how we can find the tipping points to get ahead of the curve.

Jason Clay, PhD, Senior Vice President of Market Transformation, WWF

This book tells an inspiring story about a just-started journey to a more inclusive and sustainable economy. More than enough reasons to continue.

Nico Roozen, Director of Solidaridad Network and Founder of Max Havelaar

Sustainability of global agricultural commodities is at a critical tipping point. We are at the junction between fragmented niche efforts that are falling short, and collective systematic action that realizes breakthrough impact with farmers and supply chains. This book maps the transformation and offers a path forwards. A must-read for leaders interested in long-lasting impact at sector-wide scale.

Barry Parkin, Chief Sustainability Officer, Mars Incorporated

Governments, NGOs, and industries alike have known for a long time that our agricultural systems are not sustainable; deeply rooted behaviors and systems by all parties need to change. Making changes that will fundamentally reshape the sector requires a long-term vision on how to do business in a sustainable way. Lucas Simons' book is compelling in that it provides an actionable framework to enable the change, stimulates all parties to think beyond their natural boundaries, and gives hope and certainty that sustainable change is within reach.

Isabelle C.H. Esser, Senior Vice President Foods R&D, Unilever

Changing the Food Game provides an in-depth spotlight on the complexities, struggles, and transformations within commodity chains. An interesting read.

Bill Guyton, President of the World Cocoa Foundation

Readable, interesting and relevant... it has a valuable contribution to make.

Bruno Dyck, Asper School of Business, University of Manitoba

Food security is one of the greatest challenges of our time. This book clearly shows how business can contribute to the solution, providing valuable insights and a great read.

Feike Sijbesma, CEO, Royal DSM

The way we deal with our global challenges has changed significantly over the past 10–20 years. We are much more focused on scale. Currently, unusual coalitions of competitors and key stakeholders are working together in new ways to bring systemic and structural change to the way we do business. This book gives a compelling framework to initiate, facilitate, and accelerate complex change processes. A must-read!

Joost Oorthuizen, CEO, The Sustainable Trade Initiative

Changing the Food Game

Market Transformation Strategies for Sustainable Agriculture

CHANGING THE FOOD GAME

MARKET TRANSFORMATION STRATEGIES FOR SUSTAINABLE AGRICULTURE

LUCAS SIMONS

LONDON AND NEW YORK

First published 2015 by Greenleaf Publishing Limited

Published 2017 by Routledge
2 Park Square, Milton Park, Abingdon, Oxon OX14 4RN
711 Third Avenue, New York, NY 10017, USA

Routledge is an imprint of the Taylor & Francis Group, an informa business

Cover by Arianna Osti (ariannaosti.com)

British Library Cataloguing in Publication Data:
A catalogue record for this book is available from the British Library.

ISBN-13: 978-1-78353-231-5 [hbk]
ISBN-13: 978-1-78353-230-8 [pbk]

Contents

Preface

We live in fascinating times. Never before has our world population grown so fast. Never before have so many people had so much money to spend on consumption. Never before has the projected demand for more resources, energy, raw materials, and food been so high. And never before have our global economies destroyed so much of our natural resources and ecosystems, leaving hundreds of millions of people behind in poverty.

We read about these issues in the newspaper while we eat our breakfast and drink our coffee or tea. We read the news flashes on our phones during lunch and watch televised reports on the evening news while we eat our dinner with a glass of red wine. We do this without realizing that this same breakfast, lunch, and dinner are perhaps the biggest drivers of poverty, social abuse, and environmental degradation of all. Of all global sectors, agriculture is probably the most unsustainable. And to feed our growing population, predictions are that, in the next 40 years, we will have to almost double our food production. Fixing agriculture is probably the challenge of our generation and we will not get a second chance to get it right.

While this may sound grim and ominous, I have written this book with a positive message. A book about large-scale systemic change, new ways of doing business, new partnerships, and new strategies to change the way food is produced. This is not just an idea or a plan, it is actually happening on a large scale in many sectors. Change is coming and it is important that you know this as it will not only influence your future, your children's future, but also that of the organization or the industry you work in.

For many years I have worked in the field of sustainable market transformation. This means that I take a macro, systemic view on why complex problems emerge and persist, what are the drivers behind it and look for the systemic pressure points that, when pushed, can change the system. It also means that I do not believe in quick wins and easy solutions. Instead, I ask questions such as: Why does unsustainability seem to be the natural outcome of our collective actions in the marketplace? What drives individual behavior in a larger system and what can be done to change that? I work primarily in global agricultural systems for three reasons: the need is highest in this sector; it affects the most people; and it is the most interesting, fascinating, real, uplifting, and sobering sector of all. It doesn't get any more real than when you are dealing with farmers in Africa or other emerging markets.

The topic of market transformation and systemic change processes in agriculture may sound daunting. Perhaps you are thinking, that sounds interesting, but I don't know anything about agriculture or systemic change processes. Then this book is especially for you. It explains, in simple terms, why agricultural markets across the globe are—and continue to be—so unsustainable, and describes which systemic change processes are currently happening in many agricultural markets. These change processes are described from a practical, easy, third-person viewpoint. The book deliberately avoids being too technical, too scientific, and too detailed. As a result, you will get an overview of systemic dynamics and start to see and recognize patterns of behavior, and you will understand generic intervention strategies that eventually will lead to major systemic change in the sector. And these patterns and dynamics can be applied to other major challenges as well.

Acknowledgements

Changing markets and systems cannot be done alone; it is always a partnership effort. The knowledge and insights in this book were developed over time only because I had the honor and pleasure to work with many devoted, inspiring, and professional colleagues, leaders, and organizations from all over the world, from industry, trade, government, nongovernmental organizations (NGOs), farmer organizations, multilaterals, research centers, and consultancy firms. Moreover, I have the honor to be one of the World Economic Forum's Young Global Leaders and a member of the Ashoka Fellows Network: both are sources of real inspiration, hope, and learnings. In this book I pay tribute to some of these great organizations and individuals, but I had to leave out so many others. For this I apologize. I wish society could give more recognition to these heroes who wake up and stand up for a better world every day.

When writing this book I had the privilege of being able to count on many professional people who have helped me with the writing, research, and interviews, and who have given me valuable feedback, even at times when I really didn't want any! Without their help, this book would not have been possible. I would like to briefly acknowledge them here. Jorrit Reintjes, who did much of the research, many of the interviews and prepared the basic outline of the book. Suzanne Uittenbogaard, who compiled the detailed sector fact sheets and reviewed the entire manuscript. Timo van Dun, Migle Damaskaite and Sarah Muller, who provided project assistance. Sharon Hesp, who critically reviewed the book. Silvana Paniagua, who did the graphic and creative design. Merel Segers, who helped with co-writing and editorial support, and Charles Frink, who did the final editing.

I also would like to acknowledge all the people we interviewed and whose thoughts, references, and wisdom have enriched and substantiated the book. The content of the book is very much an acknowledgement of the great work they are all doing. However, they do not necessarily endorse the content of this book; the opinions, statements, and general message of this book, about what drives systemic failure and how that can eventually be changed, are my own. All the people who were interviewed are listed in the bibliography.

I wish to thank the members of the editorial board who have spent their valuable time reading the manuscript and giving me honest, unpolished input and feedback. Sometimes straight in the face, but the book has improved because of it. The editorial board members are: Edna Kissmann (one of the owners and senior partner in Kissmann Langford), Felix Oldenburg (Director of Ashoka Germany and Europe), Gerda van Dijk (Director of the Board at the Zijlstra Center for Public Control and Governance and Professor of Organizational Ecology at the University of Tilburg), Lucian Peppelenbos (Director of Learning and Innovation at IDH) and Shatadru Chattopadhayah (Managing Director of Solidaridad South & South-East Asia), Bruno Dyck (Professor at the University of Manitoba and second expert editor), and Anne Schoemaker (Management Team SCOPEinsight). And, of course, I wish to thank Rebecca Macklin from Greenleaf Publishing. Greenleaf Publishing believed in the topic of sustainable market transformation in agriculture from the beginning and in me as the author, and I thank her and the team for their continuous support, ideas, and encouragement.

A very special word of recognition and gratitude goes to Nick Bocklandt, the Belgian coffee grower from Guatemala and co-founder of the Utz Kapeh Foundation. Nick is the hero in my personal story. This story began with him as he set me on the path I continue to walk today. His dedication, care, and loyalty for the Mayan Indian people made a deep impact on me. When times get tough I still repeat his mantra "Happy-Healthy-Strong."

Most of all I wish to thank Pauline Simons, my partner, for her continuous love, support, endurance, and patience. Besides her own busy and impactful life as a fantastic mother, teacher, and student, she had to put up with me travelling around the world, managing two companies, and writing a book. I could have done none of this without her.

This book is dedicated to my children Carmen (9), Vincent (7), and Lizet (6). They continue to be my source of unconditional love, joy, life energy, wonder, and inspiration. It is with their future and their children's future in mind that I tell this story.

How to read this book

The title of this book, "Changing the Food Game: Market Transformation Strategies for Sustainable Agriculture," suggests that systemic change is possible. There is even a strategy, a formula, and a model that can help you recognize the patterns and phases of change, and that can help guide leaders to initiate, accelerate, and drive systemic change and solve some of the complex problems of our generation. Solving complex problems may not be easy, it cannot be done quickly, but systemic change is possible. It is happening all around you, you just need to be able to see it, recognize it, and have the willingness to act or support those that do. If you want to know how you can take action as well or offer support, then read the final chapter of this book as it gives examples of great leaders who have decided to take action rather than wait and it mentions different organizations that do fantastic and important work in supporting them.

The book is in two parts. Part I consists of Chapters 1 to 4. It describes the global challenge, and shows the importance of our global food producing systems and the enormous impact it has on our economies, ecologies, and societies. It also explains what goes wrong in the system that drives this negative impact. You will come to see why our global food producing system serves as such a perfect example to explain the causes and drivers of failing systems. Once you understand what drives the race to the bottom in global agriculture, and how this is the result of our collective actions, you will start to see and recognize similar patterns causing other large and complex societal problems as well. With this understanding, we can start thinking about how to change the rules of the game in such a way that the system becomes more sustainable.

Part II consists of Chapters 5 to 10. These chapters introduce an approach to initiate and accelerate systemic change. Chapter 5 is particularly important because, and I encourage you to read it carefully, it introduces the theory of change for dealing with systemic drivers that cause the problem in the first place. Changing failing systems is all about changing the rules of the game to start rewarding the right behavior. This is done by creating a higher level of awareness in the actors in the system and a higher level of connectability within the sector. Systemic change processes do not happen suddenly: they go through various sequential phases. Each phase has its own tactics, agendas, and partnerships, and if you want to be a systemic change agent yourself or support them, it is important that you recognize the different phases and understand "what to do," "how to do it," "with whom to do it," and "how you know when it is the right time to switch gears to the next phase." If done well, it is possible to initiate, accelerate, and influence a systemic change process. If done badly, it is possible to slow down or stop a change process, or even cause it to fail. So let's get it right, as I believe we no longer have any time left to slow down or fail to implement the necessary change.

Chapters 6 through 10 in Part II contain real-life examples of agricultural commodity markets that have gone through, or are currently going through, these phases of systemic change.

My aim with this book is to help you see what drives so many of our systemic and complex problems in society. I want you to understand and see that almost all of them are the result of the same principles, the same patterns and same kind of (wrong) incentives that lead to the wrong collective actions that ultimately drive systemic failure. I hope that this book makes you optimistic about the opportunities for positive change on a large scale. Moreover, after reading this book, I hope that you will decide to join us on the exciting journey toward sustainable market transformation. In the famous words of renowned anthropologist Margaret Mead:

> "Never doubt that a small group of thoughtful, committed citizens can change the world; indeed, it's the only thing that ever has."

PART I

What is the problem?

A systemic analysis of why agricultural markets fail

1
Guatemala, where it all began

I had been sitting in a four-wheel-drive car for hours, holding on to the door while we maneuvered abruptly around potholes and sharp curves. We had left the main road long ago and were driving on a dirt road. It was September 2002 and I was sitting in the car with Nick Bocklandt, a Belgian coffee farmer who had been living in Guatemala for more than 20 years. Nick knew the road by heart and drove the dirt roads confidently at high speed. We drove through beautiful landscapes, hilltops with spectacular views, forested areas, waterfalls, plots of land with maize, beans and chickens, forgotten villages, and isolated homes.

After an eight-hour drive from Guatemala City, there, at the absolute end of the road, we arrived at El Volcán, a mid-sized coffee farm in the Guatemalan Highlands. I got out of the car, still a bit shaken from the long, rough drive, and looked around. In front of me was an astonishing view over the hills. Behind me I heard some laughter and voices. When I turned around, I saw a group of about ten Mayan Indian women wearing beautiful traditional dresses looking, pointing at me, and laughing. I didn't quite know how to behave, so I smiled and mumbled a simple *buenos dias*, and walked away. It must have been quite a sight for them to see a sunburnt gringo with a bald head, wearing sunglasses, very out of place.

This was my very first encounter with Guatemala, with a coffee farm, and with Mayan Indian women. Two months earlier I had been hired as the manager of an organization called Utz Kapeh, which means "good coffee" in Q'eqchi, the Mayan dialect spoken in that part of Guatemala. At the time, Utz Kapeh was a new initiative with big and bold ambitions to make the global coffee sector more sustainable by introducing a global standard for sustainable coffee production.[1] The idea was to work with the major coffee brands and coffee roasting companies. The idea was that these large coffee brands would use the Utz Kapeh standard as an instrument to source sustainable coffee and use the claim of sustainability assurance in their brand and product marketing. As the new manager, I was there on my first learning trip, and I got what I came for.

The year 2002 was the height of the big coffee crisis. For more than two years global coffee prices had been steadily declining and had fallen to their lowest levels in history. In the years preceding the coffee crisis there was a constant but significant increase in the supply of coffee, mainly from Vietnam.[2] With major support from the World Bank, Vietnam had rapidly developed into the second largest coffee producing country in the world. Moreover, Brazil, which is the largest coffee producer, had also increased its production. With both countries producing record volumes of coffee, world prices fell from well over $1 per pound of green coffee to as low as $0.40 per pound.

The Guatemalan coffee sector was (and still is) known and praised for its specialty: mountain-grown coffee with a full body, pleasant acidity, and a delicate, sweet aroma. Coffee growing normally provided income to about 90,000 farmers and their households.[3,4] With prices plummeting, however, they fell into poverty and despair. The price collapse was a disaster for all of the Central American regional economies, causing some 600,000 people to flee the countryside and move to the cities, leaving everything behind. In total, 1.2 million people in the region required direct food aid.[5] People were desperate, cutting down trees and occupying plots of land everywhere to grow something they could eat in order to survive.

The coffee crisis was the result of what economists call the "hogs-cycle" theory.[6] But what I witnessed was not theory at all, but real people getting into real problems on a massive scale.

In the midst of that crisis some coffee traders continued to bargain down prices even further, thus exploiting the despair of the farmers who were trying to survive. The government of Guatemala was struggling: it was unprepared, unequipped, perhaps even unwilling, to handle the situation and give people the proper support and basic means to cope with disaster.

It was during this journey in Guatemala that I saw and really understood for the first time what happens in a global economy when people are only interested in "what's in it for them" and ignore the longer term effects of their collective behavior.

Nick Bocklandt was different. At that time Nick owned several coffee estates, and was responsible for the fate of hundreds of Mayan families living on the coffee farms as workers. Born in Belgium, he came to Guatemala in 1984, bringing his European point of view with him. When he became the owner of the coffee farms, he did something that was outrageous at the time. He felt personally responsible for the Mayan people who lived on the coffee estates, and he worked together with them to understand their needs and improve their lives. There, at the end of the road, on a coffee farm in the middle of trackless wilderness, Nick built a school for the workers' children, who otherwise would never receive any form of education. There was a small hospital, a church, a community center, barracks for the seasonal workers, and wells for drinking water. Lush forests, which had been cut down for firewood years ago, were replanted again. The workers and their families were also given small plots of land to grow their own corn and beans. It was not heaven on Earth compared to our Western standards, but these people were clearly better off. Surprisingly, Nick began these activities in the early 1980s, at a time when almost nobody cared about workers on Guatemalan coffee estates or about the environment. Nick wryly observed: "It was so unique that they even thought I was a communist."[7]

Little did he know at that time that his activities would ultimately result in a meeting with Ward de Groote (at that time the CEO of Ahold Coffee Company in the Netherlands). Later, around 2000, they joined forces to start the Utz Kapeh Foundation, an organization whose mission was to implement a global standard for responsible coffee production. Together with the great work of other sustainability standards such as Fairtrade, the Rainforest Alliance, and the Common Code for the Coffee Community (4C), Utz Kapeh ultimately grew into one of the largest certification programs for sustainable coffee, cocoa, and tea in the world and part of a global movement that would eventually change the face of agricultural commodities.

Today, there are literally hundreds of global and local sustainability standards and product labels in the marketplace, in almost every commodity sector—tea, cocoa, cut flowers, spices, soy, fish farming, tropical timber, sugar, beef and many more. You can see them when you do your daily shopping. Standards, certification programs, and product labels have proven to be a very effective instrument of change in agriculture because they reward companies who care about where they source their products, and they make companies compete on sustainability. The lives of millions of farmers and producers have been positively affected as a result of the success of these standards. However, as we will see, the use of standards is an instrument that is very effective in a certain phase of market transformation. They also have their limitations. As the change process evolves to the next phase of market transformation, other and more holistic strategies are needed to complete the market transformation cycle.

During these years at Utz Kapeh, which later became UTZ Certified, I developed a passion to understand why markets, particularly agricultural markets, become so unsustainable, and why it is so hard to change them. When you talk to most organizations and individuals working in these agricultural sectors, each of them understand very well that the larger system they work in is unsustainable and that this situation cannot continue to go on, but then they raise their shoulders, say it is not their fault, they cannot change it and even resist change nevertheless. Why is this? How can this resistance be overcome?

This is not just a small problem that happens somewhere far away in an obscure African or Latin America country. It is a global problem that will affect you and your children on the largest scale possible. Global agriculture, and our global food producing systems, are probably the most important, most critical and most unsustainable systems we have. As you will read in this book, the way we produce and trade our food has become a classic example of failing systems on a massive scale, with unprecedented implications for hundreds of millions (in reality more than a billion) of people, for many economies, and for our planet as a whole.

In the year 2050, experts predict that we will have approximately 10 billion mouths to feed (almost 3 billion more than we have today), which means that food production will have to almost double. More food will need to be produced in the next 40 years than in the last 6,000 years combined. Growing enough food for all 10 billion mouths in a sustainable way is one of the biggest challenges of our generation. We will not get a second chance to get it right. To meet this challenge, I believe that we have to fundamentally change the agricultural and food producing systems, and that we need to do this on a massive scale. And this requires systemic change through a process called market transformation.

Luckily, the seriousness of the challenge is starting to sink in. Many important people and powerful organizations are working very hard and passionately to face this challenge, with increasing success. We have already come a long way and systemic change is slowly under way. The strategies that are used to drive this systemic change are important for everyone to understand, because these strategies will largely determine your future and your children's future.

2
What you eat impacts the world

"And before you finish eating breakfast in the morning, you've depended on more than half the world."

It really boils down to this: that all life is interrelated. We are all caught in an inescapable network of mutuality, tied into a single garment of destiny. Whatever affects one directly, affects all indirectly ... Did you ever stop to think that you can't leave for your job in the morning without being dependent on most of the world? ... When you go into the kitchen to drink your coffee for the morning, that's poured into your cup by a South American. And maybe you want tea: that's poured into your cup by a Chinese. Or maybe you're desirous of having cocoa for breakfast, and that's poured into your cup by a West African. And then you reach over for your toast, and that's given to you at the hands of an English-speaking farmer, not to mention the baker. And before you finish eating breakfast in the morning, you've depended on more than half the world. This is the way our universe is structured; this is its interrelated quality. We aren't going to have peace on Earth until we recognize this basic fact of the interrelated structure of all reality.[8] (Taken from "A Christmas Sermon on Peace," Dr. Martin Luther King, 1967)

2.1 Agriculture: an economic force to be reckoned with

Food. We all need it every day. Food is our source of energy, our fuel. Food is life! Every human being depends on a nutritious and varied diet in order to survive and flourish. The consumption of food is also an essential element of our social interactions, and many cultural and religious traditions have evolved around food. Just think of the American Thanksgiving dinner, Japanese tea ceremonies, or the sharing of bread in Catholic churches. Food enriches our lives in many ways. The saying "The way to a man's heart is through his stomach" even implies that you can cook your way into your lover's heart.

Our food, the fibers for our clothes, and the raw materials for our products, all come from our lands and seas. Food is part of the primary economy, the term classical economics textbooks use for activities such as harvesting, retrieval, or extraction of raw materials from the Earth. The primary economy drives the secondary economy—the transformation of raw materials into finished goods—and the tertiary economy, which provides services to the general population as well as to businesses. From this simplified breakdown, it is clear that almost all activities in the secondary and tertiary economies derive from, and depend on, the availability of inputs from the primary economy. Throughout this book I talk mostly about agriculture, but keep in mind that other activities in the primary sector like fishery, mining, and forestry are very similar in terms of importance, dynamics, impact, and challenges.

The primary sector is, indeed, the foundation of all other economic activities and is the primary source of wealth. Over 1 billion people work in agriculture, or in the processing, trading, or manufacturing of food products, making it the second greatest source of employment worldwide.[9] There are about 525 million farmers in the world, of which 85% are smallholders: people who own plots of land smaller than 2 hectares.[10] Next to farming on land, the seas are also "farmed." Another 500 million people work in fisheries and aquaculture.[11] In most African countries, 40–80% of the population depends on producing and exporting food to make a living.[12] Let's take Kenya as an example. Its agricultural sector employs more than 75% of the workforce. In a population of approximately 41 million people, this amounts to almost 31 million people who do not work in offices, factories, or shops, but are laboring in fields every day.[13] The same is true for most, if not all, Asian and Latin American countries.[14]

If we look at the more developed economies, the percentage of GDP and proportion of people working directly in agriculture or the processing of food is far lower, but the sector is still an important driver of the economy. For example, in the Netherlands only around 3% of Dutch people work in agriculture, but the Netherlands is the second largest (re) exporter of agricultural products in the world.[15] This is why the sector plays such a vital role in the Dutch economy.

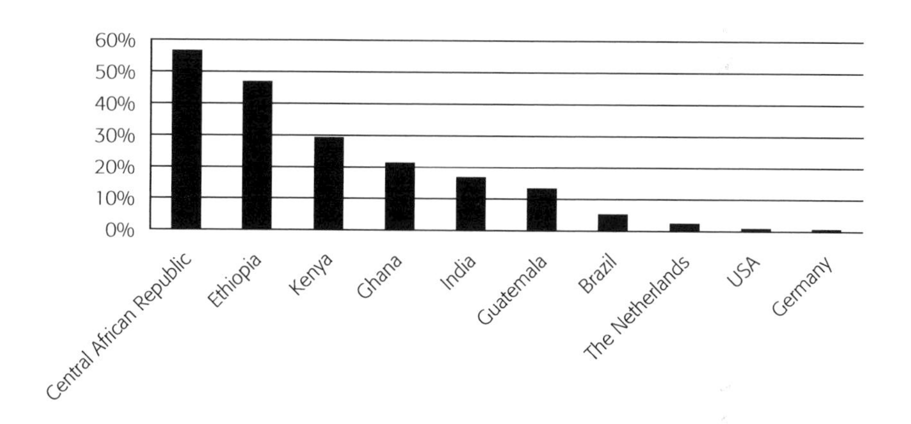

FIGURE 1 Share of the agricultural sector in countries' GDP
Source: CIA Factbook (2014)

2.1.1 Land use for food production

Agriculture is not only a very strong economic force to be reckoned with, it also has a huge effect on our landscape. If you fly in an airplane you can clearly see the impact of agriculture on the land. It doesn't matter in which direction or in what time zone you are flying, agriculture has changed the surface of the Earth profoundly. There is hardly an area left without the spotted and colored pattern caused by the production of various crops. At times the farms are neatly ordered and organized, at others they are scattered and located in forests and other kinds of natural surroundings. Even in the desert you see irrigated green circles in vast seas of sand. The statistics are even more astonishing. The total land area of the world land surface is around 130 million km^2. Agricultural land takes up almost 38% of that area.[16] While more than half of the global population lives in cities, the area used to grow our food is 60 times larger than that of all cities combined.[17] No other sector in the world uses more land than agriculture. If we include forestry and fisheries in the equation, the space needed for our global food system is even more impressive. Arable land is getting scarcer every year. Our growing population needs more land for growing cities, infrastructure, and recreational space. At the same time we lose a lot of land. According to the UN we are losing approximately 12 million hectares of land every year (23 hectares/minute) due to desertification and soil degradation. On this area of land we could grow 20 million tons of grain every year.[18] All in all, not that much arable land is still available to produce the increasing amount of food we need to satisfy our global hunger of our growing population. Unless we significantly change the way food is produced, we simply do not have enough available land to feed everyone.

2.1.2 Agriculture's thirst

Besides being one of the main contributors to employment and economic development in many countries, agriculture also has a severe negative impact on our environment. Take water consumption for example: 70% of all the freshwater in the world is used to irrigate crops. A thirsty individual might drink four liters of water daily, but this is nothing compared to the "hidden" water that we consume. It takes 2,000 to 5,000 liters of water to grow and process the food that you consume in one day.[19] For instance, it takes around 140 liters of water to produce a single cup of coffee.[20] This includes the water used to grow the coffee plants, process the beans and transport them to you. Together, all coffee drinkers indirectly consume about 110,000 billion liters of water to satisfy their yearly coffee needs.[21] This seems like a lot, but it is nothing compared to the water used to produce meat. It takes 2,393 liters of water to produce a single hamburger, and a kilo of steak requires 15,415 liters of water.[22]

By 2025, water withdrawals are predicted to increase by 50% in developing countries and 18% in developed countries.[23] This is a serious problem because, in many parts of the world, water is becoming scarce.[24] The Aral Sea (on the border of Uzbekistan and Kazakhstan) is an example of how the huge need for agricultural water impacts water basins. In the 1960s, the two main rivers bringing water to the Aral Sea were diverted to irrigate cotton fields. The sea shrank to a mere 10% of its original volume, which caused increased concentrations of contaminants and salt. This had a devastating effect on the whole region: it destroyed the fishing industry, changed the local climate, impacted people's health, and damaged other agricultural crops.[25] Another example is Lake Naivasha, one of the few freshwater lakes in Kenya. Water from the lake is used to grow flowers. The water level in the lake has dropped 4 meters since the first flower farms started production in the early 1980s.[26] Water problems have become more visible in other regions as well. In 2013, the Indian government revealed that residents in 22 out of 32 major cities and regions have to deal with daily water shortages.[27] The drop in groundwater is especially severe in parts of California, India, the Middle East, China, and large parts of Africa. The main cause is increased water demand from agriculture.[28]

2.1.3 Agriculture versus biodiversity

Biodiversity loss is probably not the first thing that comes to mind when eating a juicy steak or sipping an espresso, yet the two are very much related.[29,30] Agriculture is the number one reason for loss of biodiversity worldwide. Several factors play a role here.

Forests and other natural habitats are disappearing rapidly as a consequence of agricultural practices.[31] Agriculture is the largest driver of all for deforestation, and slashing and burning is an ancient method used frequently by farmers to expand or replace their arable land; it is an easy, cheap, and quick way to both clear the land and fertilize it at the same time with the charred remains of the forests. The downside of this practice is that the fertility of the soil diminishes quickly, as the topsoil in forests is often very thin. In many cases, production falls after only a few years, and this means that new plots of forests have to be burned again and again. Slashing and burning is not only often used by individual smallholder farmers, it also happens on a larger scale. Brazil, for instance, was the absolute world leader in terms of total deforested area between 1990 and 2005. The country lost 42,330,000 hectares of forest—roughly the size of California.[32] The soybean and the cattle industries take the lead here. Both sectors need vast amounts of land and are moving into the rainforest in Latin America, Africa, and Asia because land is less expensive there. In order to convert forest to farmland, they simply burn it down. When the soil is exhausted, they move on.[33] For the Amazon as a whole, which stretches far over the borders of Brazil into Peru, Bolivia, Colombia, Ecuador, Guyana, Suriname, and Venezuela. In the last 40 years, nearly 20% of the Amazon rainforest has been cut down and burned.[34] Palm oil is also no stranger to deforestation and burning practices. These days, we can hardly do without palm oil; it is an important ingredient in almost 60% of all the products we buy in supermarkets (such as cookies, shampoo, noodles, soap, cosmetics, chocolate, ice cream, and many more). Between 2009 and 2011, palm oil was the largest driver of deforestation in Indonesia. In September 2013, Greenpeace calculated that palm oil production accounted for about a quarter of the country's forest loss.[35]

Another important driver for loss of life is the use of large amounts of chemicals. Agriculture is one of the biggest consumers of chemicals in the form of fertilizers, herbicides, and pesticides, many of which are extremely harmful to the environment.[36] For example, the half-life of pesticides (which is the amount of time required for a quantity of the chemicals to fall to half of its "toxic" value compared to the beginning of the time period) is often between one and four years. This means that, even if pesticides are sprayed on the fields just once a year (they are used much more often in reality), they will accumulate. This makes the land sterile and pollutes groundwater. Insects die, which in turn affects the bird and mammal populations. The use of pesticides is a major reason why various species are threatened with extinction, ranging from bees to butterflies, snakes, and condors, frogs, toads, and dolphins.[37,38,39]

Another important driver of loss of biodiversity related to food production is eutrophication. Eutrophication is the process by which land or water bodies become enriched with (dissolved) nutrients like phosphates or nitrogen. These nutrients come from excessive use of fertilizers, livestock run-offs or use of fish feed in the water for the production of fish. Eventually they end up in ground water and water streams where they are favorable to some species like algae or grass but are affecting many species of microorganisms, insects, fish, plants and animals that are dependent on clear and clean water or that can only survive on "nutrient poor soils or waters."[40,41,42]

2.1.4 Agriculture and climate change

Carbon neutral was the "Oxford Word of the Year" in 2006.[43] Agriculture, however, is nowhere near being carbon neutral. In fact, agriculture is one of the largest contributors to climate change.[44] According to the most recent data available from the Worldwatch Institute, "global greenhouse gas emissions from the agricultural sector totaled 4.69 billion tons of carbon dioxide (CO_2) equivalent in 2010." In comparison, global CO_2 emissions from all global transports and logistics totaled 6.76 billion tons in 2010, and emissions from electricity and heat production reached 12.48 billion tons.[45] Agriculture is the third largest emitter and is directly responsible for about 14% of total global greenhouse gas emissions, while forestry, deforestation, land clearing for agricultural purposes, and fires or decay of peat soils account for another 17%.[46]

Carbon dioxide emissions in agriculture are produced in various ways. Most CO_2 emissions from agriculture result from cultivation and decomposition of organic soils (these are soils, for example, in wetlands, peat lands, bogs, or fens with high levels of organic material). Before you can grow crops on these soils, they must be drained (lowering the water level). In order to do this successfully, techniques such as drainage ditches are used to dispose of the water. As the water disappears, the organic matter is exposed to oxygen-rich air and starts to decompose, emitting large amounts of CO_2 and methane (CH_4) to the atmosphere. In total, these practices account for around 14% of total agricultural greenhouse gas emissions.[47]

Methane is an important and particularly potent greenhouse gas. One unit of CH_4 is 23 times more powerful in contributing to climate change than one unit of CO_2.[48] Methane is produced in multiple ways. Besides the decomposition of organic material, the biggest emitter is livestock. And, contrary to popular belief, most CH_4 does not come from the back end of the cow, but from the front end: the belching of these ruminant livestock. One animal generates between 250 and 500 liters of CH_4 per day; on average, cattle belch 6% of their ingested energy as CH_4.[49] Livestock farming alone produces 35% to 40% of all CH_4 emissions caused by human activity, while 50% of total agricultural greenhouse gas emissions consist of this gas.[50,51] The irony of this is that, besides being one of the biggest contributors to climate change, agriculture will be one of its biggest victims of it as well. Climate change itself is bad news for agriculture.

Increasing temperatures, combined with changing weather patterns, will have a profound impact on agriculture. The World Bank has warned about risks for a number of regions due to climate change. According to their reports, "food production is at risk in Sub-Saharan Africa, productivity and coastal zones are at risk in South East Asia and there are risks of extreme water scarcity and excess in South Asia."[52] And "climate change affects agricultural production through its effects on the timing, intensity, and variability of rainfall, shifts in temperatures and carbon dioxide concentrations."[53] Or "in the tropics and subtropics, crop yields may fall by 10% to 20% in 2050 because of warming and drying, but there are places where yield losses may be much more severe."[54] According to *The Guardian*, higher temperatures in the Middle East and northern Africa will lead to decreased yields of around 30% for rice, about 47% for maize, and 20% for wheat. Only a few crops can survive average temperature rises of more than 2°C. Brazil is already being severely affected by a temperature rise of 1–2°C. Its production of coffee, rice, beans, manioc, maize, and soy are all expected to decline.[55]

In some areas, climate change will cause more excessive rainfall. Too much rain causes topsoil to be washed away. This loss is permanent. Less topsoil means lower agricultural production, and can result in a huge drop in annual yields per hectare.[56] To make matters worse, climate change will most probably make water resources scarcer which increase the rate of desertification and the rising temperature will encourage weeds and pests, which in turn will increase the need to use pesticides and herbicides.[57]

Interestingly, not all regions are expected to be negatively affected. The International Panel on Climate Change (IPCC) has calculated that certain regions of the U.S., Russia, and Asia will become more suitable for agricultural production and so actually stand to gain from climate change.[58] But for most countries, the higher temperatures mean that it may no longer be possible to grow climate-sensitive crops in areas where they are customarily grown.

2.1.5 Agriculture: the farmers and their families

From a social point of view, agriculture leaves its mark as well. Adults and children, men and women—many are affected negatively. Child labor, slavery, hunger, and poverty are, in many cases, directly related to agricultural production.[59,60,61,62]

In developing countries, many workers are self-employed in precarious conditions, or are employed on a casual basis without either a contract or access to social security (this is called informal employment). In most cases, informal employment means lower, more variable, pay and poorer working conditions than formal employment. This informality is highly prevalent in sectors such as agriculture and services, although it is not limited to these sectors.

The vulnerable employment rate is a newly defined measure of persons who are employed under relatively precarious circumstances.[63] Vulnerable employment in Sub-Saharan Africa was estimated at 77.4% in 2013, which is the highest rate of all regions. There is also a connection between vulnerable employment and poverty: if the proportion of vulnerable workers is sizeable, it may be an indication of widespread poverty. The connection arises because workers with vulnerable employment status lack the social protection and safety nets to guard against times of low economic demand, and often are incapable of generating sufficient savings for themselves and their families to offset these times.

Agriculture is among the most hazardous sectors and occupations.[64] Occupational deaths and injuries, and work-related diseases, take a particularly heavy toll in developing countries, where large numbers of workers are concentrated in the primary and extractive activities. Most common and frequent occurring injuries are accidents to cause loss of limbs, pesticide exposure, pulmonary disease and musculoskeletal disorders.[65]

Moreover, agricultural production in developing countries is known for child and forced labor. It is estimated that over 129 million girls and boys, aged between 5 to 17 years, work in agriculture. This means that 60% of all child laborers work in agricultural production. They often do dangerous or very strenuous physical work.[66] Because these children are working long hours on the land, they can't go to school, so their prospects for a different future are limited.

While most people associate slavery with the slave trade of previous centuries, it remains an important issue in agriculture today. According to the International Labour Organization (ILO): "almost 21 million people are victims of forced labor in the production of food crops or fiber—11.4 million of them are women and girls, and 9.5 million are men and boys."[67] Other sources speak of almost 30 million enslaved people.[68] Depending on the source this means that, out of every 1,000 people, three or four are modern-day slaves. Some numbers from the Brazilian agricultural sector highlight the problem: 25,000 to 100,000 people are enslaved, and almost all of them do agricultural work. Slave labor is used in Brazil to harvest sugarcane, to clear vast amounts of land to raise cattle, and to provide access to valuable timber.[69]

Ironically, undernourishment—an intake of food lower than the minimum needed for good health and growth—is also a major problem in agriculture. Besides child labor and forced labor, many rural areas are also characterized by high levels of stunting and undernourishment. According to Unicef: "Three-quarters of all hungry people in the world live in rural areas. In six countries of East and South Asia at least half of children under the age of 5 are stunted, as are 40% of children in Sub-Saharan Africa."[70]

We have reached the paradoxical situation where the people who are most dependent on agriculture and growing our food for a living are the ones who get the least food, while the people who are consuming the food are growing fatter every day.[71] According to the Food and Agriculture Organization (FAO), worldwide obesity has increased two-fold since 1980 and it has reached epidemic proportions globally, with at least 2.8 million people dying each year as a result of being overweight or obese.[72] What a world we live in.

2.2 Feeding our growing population

Clearly, the way we produce food is out of control and is cause for deep concern. And the challenge is about to become a whole lot bigger. The catch is that we will need to almost double our food production in the next 40 years.

Our booming global population is an undeniable fact. It exploded over the last few hundred years and will keep on growing for several decades to come. Because of the decreasing mortality rate of children, the growing percentage of women surviving childbirth, and a significant increase in life expectancy, the world population is estimated to keep on growing to around 9.6 billion people by 2050 (Fig. 2) and around 10.9 billion by 2100.[73] This means that, in the next 36 years, the world will have 2.4 billion additional mouths to feed. And these mouths will be hungry and demanding indeed.

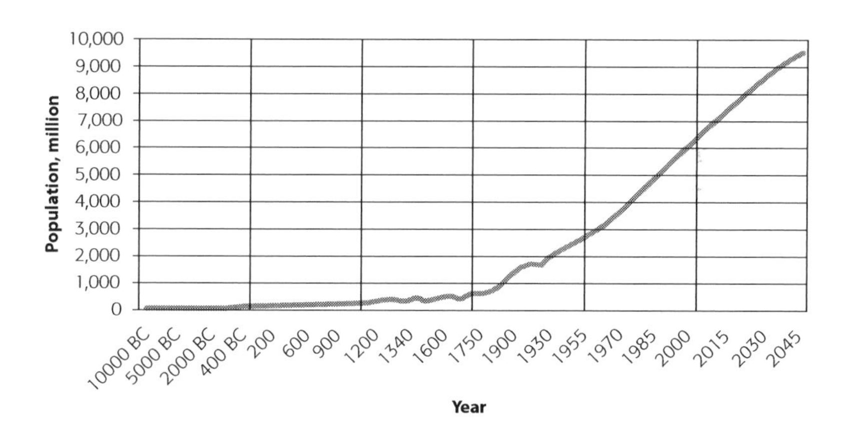

FIGURE 2 Global population growth[74,75]

Population growth rates are not equally distributed around the world. Most developed countries will see only small changes in their populations, while less-developed regions will account for almost all population growth. Today, China has the largest population in the world, but India's population will surpass China's within the next 14 years.[76] Africa has one of the fastest-growing populations. The populations of Burundi, Malawi, Mali, Niger, Nigeria, Somalia, Uganda, Tanzania, and Zambia are projected to increase at least five-fold by 2100. In fact, today in Nigeria alone more babies are born every year than in all of Europe.[77] Africa's population is expected to double between 2013 and 2050 and double again between 2050 and 2100. In 2100, Africa's population is predicted to total 4.2 billion people, which will be more than 38% of the world population at that time.[78] These growth figures should be a major cause of concern for everybody.

Feeding 10 billion mouths is one thing. The real challenge is that these mouths will be more demanding than ever. During the same period of time in which we will reach a population of 10 billion people, global GDP is expected to triple as well.[79] As a result we will see a spectacular growth of middle-class incomes in the coming decades.[80] This means that more people than ever will have more money to spend than ever. When people have money to spend, they will want to have more, better, and more varied food than their parents and grandparents used to eat. In particular, the demand for animal protein (meat and fish) will increase.[81] As you can imagine, raising chickens, cows, or pigs is not the most efficient way of producing food because most of the calories in animal feed are metabolized by the animal itself.[82] In some cases it takes between 6 and 10 kilos of animal feed to produce 1 kilo of animal protein. It is this combination of factors that will lead to a real explosion in food demand in the next decades.

The increasing use of biofuels is another important driver of the demand for more food production.[83] Biofuels are becoming an increasing part of the global energy mix. The main sources for biofuels are corn, soy, sugarcane, rapeseed, palm oil, and wheat.[84] As an example, we can look at corn and soybeans in the U.S. In 2010–2011, 40% of the total corn crop and 14% of soybean oil production was used to produce biofuels and other products, including distiller's grains for use as animal feed.[85] This adds to the already stressed-out global agricultural system.

2.2.1 The biggest challenge of our generation

To cover the increasing demand for more food, the UN predicts that we will need to raise overall food production by 70% before 2050.[86] If these projections are right, then we will need to produce more food for human consumption in the next 40 years than all the food produced in the last 6,000 years combined![87] Of course not all experts agree on these numbers. Some point out that we need to do more than increase it by 70%; that we actually have to double it.[88] Others emphasize that we can do with a smaller increase in production if we can reduce the enormous amount of food waste throughout the food chain.

Food waste is, indeed, no small matter. One-third of all food produced worldwide ends up going to waste.[89] The causes are numerous: poor harvesting practices; inefficient storage and transport methods; as well as wasteful retailer and consumer behavior. Vast amounts of good top-quality food are thrown away by retailers, restaurants, conferences, and catering companies where food safety regulations prohibits the use of this food for consumption. In total, 1.3 billion tons of food is wasted each year. The direct economic cost of this wasteful behavior is estimated at $750 billion annually. At the same time, food waste adds 3.3 billion tons of greenhouse gases to the atmosphere.[90] Both economically and ethically it makes a lot of sense to start by reducing food waste significantly. But even if we could reduce food waste by 50%, we will still need to increase food production by approximately 50–60%.

Higher productivity (having more production per hectare) seems to be the only plausible method to meet this challenge. But we must do it in a more sustainable and inclusive way. According to a high-level FAO expert forum on how to feed the world in 2050, 90% of the growth in crop production is expected to come from higher yields and increased cropping intensity. In particular, more corn, wheat, rice, soybeans, and sugar will be needed to meet food, feed, and industrial demand.[91] Only 10% is expected to come through land expansion. There is simply not enough land to continue with a business-as-usual scenario.

2.2.2 The price of scarcity

The imminent scarcity in food resources has been a concern for companies and national governments on the highest level for the last ten years. In fact, we can see the effects of this already. Food prices are increasing across the board and becoming more volatile. The historical FAO food price index (Fig. 3) clearly shows price increases of several commodities between 1990 and 2013, reflecting the increasing scarcity.[92] In the past ten years, average prices have increased by 100–150%.

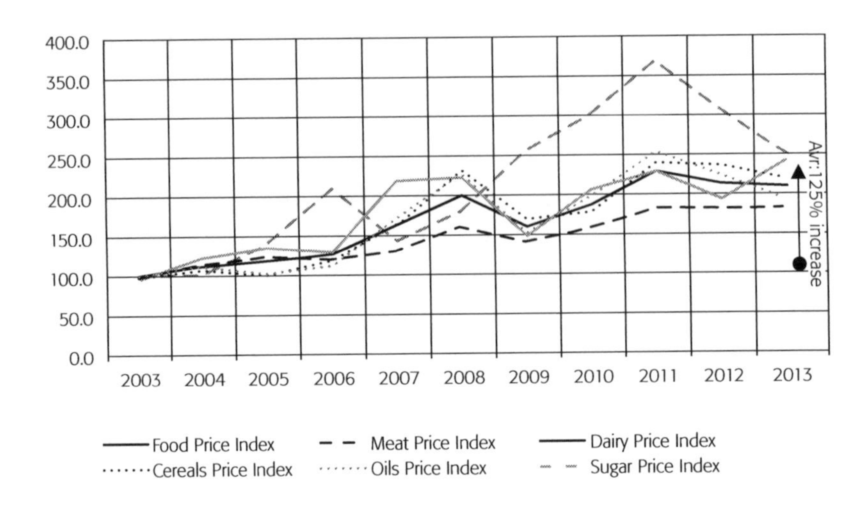

FIGURE 3 FAO food price index[93]

In most of the developed economies increasing food prices are a nuisance at worst. Most of us can afford to pay more for our bread and meat. Rising food prices in developing economies, on the other hand, are no laughing matter. Poor people in developing countries often spend between 60% and 80% of their disposable income on food.[94] An increase of the price of basic food ingredients has serious, direct effects on their livelihood and wellbeing and on the economic growth and stability of countries as a whole. A good example is the Arab Spring in Egypt. The images of brave citizens bringing down the dictatorial regime are iconic. Perhaps less well known is that the revolt was also driven by increasing food and energy prices. In 2007–2008 grain prices rose significantly, which caused the price of bread to rise by 37%. As unemployment was also increasing, more and more people depended on subsidized food. But the government of President Mubarak didn't provide more food, which stirred unrest. After the revolt in Egypt, other Arab regimes responded quickly by adjusting food prices and providing more subsidies.[95]

Perhaps the most famous example of social unrest caused by increasing food prices was Mexico's 2007 tortilla crisis, which eloquently illustrates the complex interconnectedness of our global food systems and the effect this has on real people's lives.

The Tortilla crisis in Mexico

The story began in 1993 when the U.S. had an enormous corn surplus. Drastic measures were needed to get rid of the surplus, which was a driving force behind the NAFTA treaty (North American Free Trade Agreement) of 1994. The treaty was part of the solution for the U.S. because it enabled it to dump its surplus corn on the Mexican market at very low prices. Around 2 million Mexican smallholder farmers, whose income depended entirely on corn, couldn't compete with the cheap corn and were forced to switch to other crops. As a result, corn production fell significantly in Mexico.

At the beginning of 2000, the demand for corn rose again in the U.S. because it was becoming a source for biofuels. Nowadays, 20% of the U.S. harvest is used for the production of bioethanol (this was a mere 6% in 2006). As a result, corn exports to Mexico fell sharply. As many Mexican farmers had stopped growing corn by then, this resulted in a shortage of corn, which was aggravated by rising demand due to the growing population. Consequently, corn prices skyrocketed and poor Mexicans could no longer afford to buy their traditional tortillas. This was a serious problem as tortillas provide 50% to 60% of the daily calorie intake of low-income consumers.[96] In Mexico, household expenditure on food and non-alcoholic beverages was 26.7% of total expenditure in 2008, relative to 13.1% in the EU.[97] The sharp increase in food prices therefore immediately affected spending on housing, education, healthcare, and transportation. In 2007 the situation reached boiling point when tens of thousands of Mexicans took to the streets and marched. This resulted in riots and demonstrations, and left many people injured.

I believe we will see much more of these types of civil unrest and riots as prices of food and energy will continue to rise.

2.3 We need radical, systemic change

It is a matter of simple arithmetic. If we already use 40% of our total land surface and 70% of our fresh water to produce our food, if agriculture is already the main driver for deforestation and biodiversity loss, poverty, child labor, and forced labor, then it is self-evident that we cannot simply double our agricultural output while continuing to do business as usual.[98] Something fundamental has to change. We have to do better in a radically smarter, more efficient, and more sustainable way than ever before. This must happen on the largest systemic scale imaginable. And we have less than 40 years to do it.

But before you can change a system structurally you need to understand why the system functions the way it does. You need to understand the rules of the game first before you can change them. And for this we first need to learn how to look, read, and understand systems.

3
Reading and understanding behavior in systems

The facts don't lie. Our global agricultural system is the most important, the most critical, and by far the most unsustainable system we have. The question is, of course, why is the agricultural system failing? Why is the growing and trade of the most important product on earth not leading to prosperity, development, and wellbeing, but to poverty and ecosystem destruction instead? And why can't we stop this? The answers to these questions have to do with the way we have organized our agricultural system. The underlying drivers of that system ensure that unsustainable behavior is in many cases the natural outcome.

In order to understand this better we first have to be able to look at, read, and understand systems language. This requires that we rise above a complex issue that we are trying to understand and take a "third-person viewpoint." With this third-person perspective we can see the "action–reaction patterns" that are causing many complex issues to emerge and persist.

Most movies and theatrical plays work with this third-person viewpoint. You, as the reader or viewer of the story, have oversight. This gives you the unique opportunity to not only understand the story from the character's point of view (first-person viewpoint). You also understand what other people are doing to the character and why (second-person viewpoint). You, as the viewer, however, have a unique perspective: the third-person viewpoint. You can hover above the situation and understand why the chain of action–reaction events occurs. From their individual perspectives, each character in the play would claim they acted with perfect rationality and logic, that they had no other choice but to do what they did; but the bigger drama unfolds nevertheless, even if nobody wants it to. Only you as viewer understand why the murder has taken place or why the war started.

The third-person viewpoint helps you to observe what I call the "dance of destiny." Like a dance, people act and react to each other, and this creates an emerging pattern. Understanding this action–reaction dance provides clarity into what really causes problems and therefore enables you to get out of the unhealthy "He-did-this-so-I-have-to-do-that-loop" that most of us are caught in every day.

Over the years, an extensive body of theory has developed on systems thinking, and I highly recommend reading some of these excellent books, which are mentioned in the appendix, to grasp the basic principles.[99] Important for now is to understand that the language of systems thinking is thinking in loops rather than in linear problem–solution relationships.

3.1 Thinking in loops

In systems thinking there are basically two kinds of loops: "reinforcing feedback loops" and "balancing feedback loops."

Reinforcing loops (R) cause exponential increases or decreases over time. A classic example is the self-fulfilling prophecy of a bank run, in which rumor of a bank's poor credit status causes people to withdraw money, causing a worse credit status, causing more people to withdraw their money, ultimately leading to the collapse of the bank, even if the bank was healthy to begin with. Such reinforcing effects are hard to stop within systems. Reinforcing feedback loops are sources of fast growth, explosion or erosion, and collapse in systems. Such a system will ultimately destroy itself, if there are no balancing loops.[100]

Balancing feedback loops (B) have the opposite effect: they stabilize systems over time. An example of a balancing system can be the heating system in a room that is set for 20°C. Every time the temperature drops below this threshold, the heater kicks in and warms up the room. The heater will automatically turn off when the temperature is slightly above 20°C. This pattern will continue to repeat itself while stabilizing the room temperature at around 20°C.

In most healthy systems, reinforcing loops and balancing loops coexist at the same time. Things go wrong when the reinforcing and balancing loops are no longer in balance. These feedback loops are illustrated in a simple systems diagram of a rabbit population (Fig. 4). A healthy rabbit population has both reinforcing and balancing loops, and this helps the population to remain stable and vital over time. You can probably imagine what would happen when the reinforcing or balancing loops are no longer in balance—for example, because the predators that hunt the rabbits have disappeared. It would have catastrophic implications for the rabbits, the environment, or both.

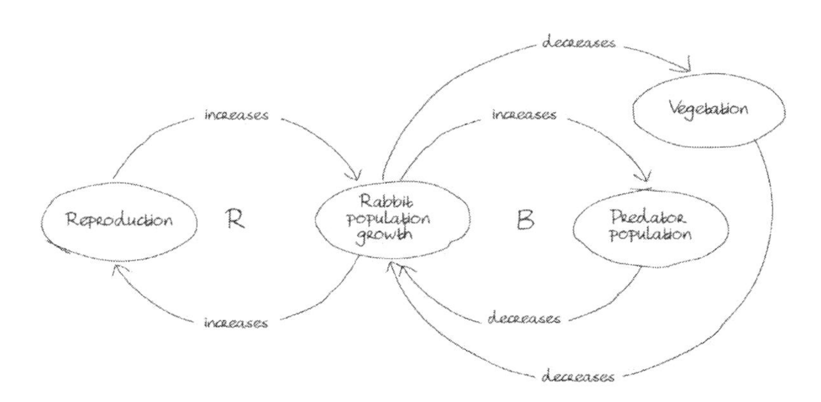

FIGURE 4 The reinforcing and balancing loops in a healthy rabbit population

Let's see these loops of "action–reaction" in action in some important examples of human system failure.

3.2 Classic examples of failing systems

"Freedom in a commons brings ruin to all." (Garrett Hardin, *The Tragedy of the Commons,* 1968)

The loops that determine the fate of a rabbit population are active in other, larger, human systems as well. In agricultural economics we find some classic theoretical models of feedback loops that, when out of balance, eventually cause whole (agricultural) markets to fail. Two famous examples are the earlier-mentioned "hogs-cycle" theory and the "Tragedy of the Commons." The fact that both theories of systems failure come from agriculture speaks volumes about what we can learn from agriculture in order to understand the drivers of human behavior and systemic dynamics and how we can fix them.

3.2.1 The hogs-cycle theory

The hogs-cycle, also known as the cobweb theorem, explains periodically recurring cycles in the production and prices of products by looking at the interaction of supply and demand.[101] Simply put, when prices for hogs (pigs) are high, farmers will invest in raising more hogs in the expectation that, by the time the hogs are ready for sale, they will benefit handsomely from the high prices. Since most of the rational, profit-seeking, hog breeders think this way, the supply of hogs increases all at the same time. The market will be flooded, prices collapse, and everyone is worse off. The lower prices cause many farmers to move out of raising hogs and start growing other crops. This leads once again to increasing hog prices and probably lower prices for other crops. The behavioral patterns of rational individual actors in the hogs-cycle theory are a frequent and important source of destruction of many businesses, asset values, and lives. Despite the destruction and havoc, this cycle continues to happen over and over again. Somehow, most people seem to be unable to learn from the past and prevent this kind of systemic behavior and act in a more contrarian manner. This is an important lesson to bear in mind.

Another famous theory of failing (agricultural) systems is the Tragedy of the Commons.

3.2.2 The Tragedy of the Commons theory

The Tragedy of the Commons is the second classic theory of failing systems. The Tragedy of the Commons is a term introduced by the American ecologist Garrett Hardin (1915–2003).[102] Hardin describes a rural community where a piece of land is commonly used by a number of farmers to graze their cattle. As each individual farmer wants to maximize his profits, they allow more and more animals to graze on the common land. This doesn't cost the farmer anything more and gives him the benefit of more milk and meat. As long as sufficient grass, water, and space are available, this process can go on happily for quite some time. But the limits will finally be reached. At some point, the grass is no longer sufficient to feed all the cattle, the land is overgrazed, and the system collapses. The farmers have enjoyed short-term gains, but they overexploited the common grassland; in the end, all the farmers are worse off.

Although the hogs-cycle theory and the Tragedy of the Commons originated from the last century, these theories are still applicable and very real today. We frequently hear about boom-and-bust cycles and price bubbles in the markets, bankrupting companies and destroying value. Or we hear about diminishing fish stocks as a result of the fact that, for each individual fishing boat, fishing fleet, or even individual fishing nation, it is more profitable to take a larger amount of fish out of the seas now, leading to more profits in the short term. The consequence, a depleted fish population, is a common problem for later or for somebody else.

The hogs-cycle theory and the Tragedy of the Commons teach us that three (out of four, which will be presented later) important principles must be present before systemic failure can take place:

1. The system must consist of self-serving actors who seek to optimize their own short-term gains

2. This self-serving behavior negatively affects others or has an adverse impact later in time (there is a time and/or physical distance between cause and effect)

3. There is no overarching effective authority or enforcement mechanism between the actors to ensure the common or public good

The reason why these three conditions lead to systemic failure is because the individuals in the system are incentivized to put their own, short-term interests above the long-term interests of the community. This behavior has become the ruling measure of success in the system. To counter this, normally an authority (usually a government agency) that acts above and on behalf of the individual actors is needed to enforce rules, implement quality or safety standards, organize feedback mechanisms, and provide enabling conditions such as infrastructure, knowledge, and financial stimuli. When these authority systems fail, are ineffective or are not there, this further intensifies the race between the actors to seek individual short-term gain by taking as much as possible out of the public good domain as quickly as possible, before the others do this. And, as we will see, this is a powerful dynamic.

3.3 Individuals don't really have a choice

When individual actors are caught in these kinds of systems, the result is a negative race in which each individual will feel they do not have a choice but to continue to play along and try to win from the others. Even when some actors do not want to play this game of short-term profit seeking, while others do, they not only lose out on their own short-term gains, but ultimately the system still collapses. The real tragedy is that most of us will indeed act according to this dominant definition of success and its related incentive structure (money, status, and position), particularly if we are surrounded by others who do the same. In systems like this, most of the actors in this system will start to behave like takers, as Adam Grant calls them in his recent book *Give and Take.*[103] According to Grant, a Taker is someone who likes to get more than they give; they are bound to put their own interests ahead: "If I don't look out for myself first, no one will." If this attitude is the dominant culture in the system, and this defines what we call success, most people will start behaving like this as well. They will claim they have no other option but to play along and take care of their own interests as well. The result of this collective action is an accelerating race to the bottom, leading eventually to the collapse of the whole system. Since everyone is doing it, individuals feel they have no real choice but to do it as well, while nobody feels individually responsible for the end result.

These systems failure dynamics are not limited to agriculture. It is a pattern we see in most of the large, complex problems our society faces. Consider our failed attempts to reduce our dependency on fossil fuel and stop or limit climate change. In the current system, countries and companies who do not want to commit to reducing fossil fuel consumption are winning and are even rewarded for it. Let me explain. Countries or companies who choose not to reduce the use of fossil fuels not only continue to benefit from a cheap source of energy now, but the future consequences are for other generations or other countries. Refusing to change their behavior becomes even more attractive when other countries or companies are willing to take action. Then, not reducing the use of fossil fuels (let's call this the wrong behavior) is rewarded even more for several reasons.

The countries who do not act continue to enjoy cheap energy. They may even gain a competitive advantage over the countries who do act (because the other countries will have a higher cost base). Moreover, they "free-ride" on the efforts of other as the co-benefits of the reduced greenhouse gas emissions achieved by these countries, as they are enjoyed by everyone, including the countries who do not act.

This is again a classic systems failure situation. The incentives for unethical behavior, going for your own self-interest in the short term, while externalizing the consequences to others, are simply too great. No wonder nobody is keen to go first and really commit themselves to fighting climate change, for to do so means going against your own interests, so it is better if others go first.

Other examples of what happens when the "wrong definition of success" is rewarded are:

- Bankers sell "toxic" financial products for short-term windfall gains and bonuses; the risks are for the clients, and when it goes wrong public money will bail them out
- Slaughterhouses caught in price wars, taking advantage of weak control mechanisms, put horsemeat in consumer products to escape continuous price pressure

- Milk factories in China add a toxic substance to watered-down baby milk to increase the apparent protein levels, which increases their profit margins[104]
- Supermarkets provide free plastic bags, which is convenient for shoppers and will increase their sales, but eventually leads to the "plastic soup" in our oceans that is now threatening the ecosystem

Examples of such behavior, which eventually lead to systemic failure, are all around us. And they are all a result of collective behavior of actors which operate in a system that follows the same kind of negative patterns.

3.4 The fourth principle: conditions for change are not there

What characterizes many of these complex problems is not only that they are caused by incentives to go for short-term, selfish gains in a context of a failing or ineffective authority, but also that any serious attempt to solve them gets stuck in complex chicken-and-egg situations. This is because the conditions that are necessary to solve the problems and break out of the negative loops are often absent. For example, for countries to reduce fossil fuel consumption, alternative energy sources must be economically viable and easily available. These alternative energy sources only become economically viable and available when you have invested in them and scaled them in the first place. But you don't because the fossil fuels are so cheap and available, and the alternative sources are expensive. Here is where we get stuck.

Preventing food fraud, as in the cases of unlabeled horsemeat and baby milk contamination, requires markets to reward quality and higher prices throughout the value chain, but many markets are stuck in price competition and so the value chain optimizes around low quality and cheap food, which created the problem to begin with. How do you break out of that?

The paradoxical situation now occurs that, when systems fail, the conditions that are necessary to change that failing system are often not present. This makes it even more difficult to change the rules of the game and to find structural solutions to the problem.

3.5 The four principles of system failing

This is the fourth dimension of failing systems. The four principles of system failures are:

1. Markets are designed in such a way that all actors are rewarded for seeking short-term benefits
2. There is a gap/distance between their actions and the consequences of their actions
3. These markets operate in the context of failing or ineffective government systems
4. The basic conditions for change are not present, keeping the system stuck in its negative momentum.

Let's apply all of what we have discussed to our global food production systems. Let's see why these production systems get caught in negative reinforcing loops and how these global systems fail on a large scale.

4
Why do agricultural markets fail?

It continues to fascinate me to see global agricultural systems fail so drastically. On the one hand, this sector is so fundamental to human life and so important for so many economies; on the other, it keeps hundreds of millions of people in poverty and leads to massive environmental degradation.

We see these sustainability issues in almost every part of agriculture, fishing, or forestry. There are hardly exceptions. While some crops are mostly linked to massive deforestation, others are related to large-scale poverty and issues such as slavery and human trafficking, while still others are mostly linked to environmental degradation or water depletion. Although the symptoms may vary, the underlying causes and principals are the same and they are very similar to the patterns and forces we have seen in the previous chapter. They are now occurring on the largest scale imaginable.

Now let's sit back, take the third-person point of view; let's think in loops, apply the principles of failing systems to agriculture, and watch the drama of the agricultural "dance of destiny" unfold.

4.1 Pssst ... come into my store

Have you ever visited a souvenir market in an exotic country? Have you noticed that most of these little shops sell exactly the same products and artifacts? To me, this doesn't make sense. If everyone sells the same thing, then it doesn't matter that much where I buy my wooden elephant or giraffe, candle and napkin holders, or placemats. I can easily make a good deal with any vendor by simply threatening to take my business elsewhere. I am the one with the bargaining power; the vendors are competing with each other on price only to get the deal and are, in that sense, price takers.

In the trade in agricultural products a similar dynamic is present, but on a bigger scale. Agricultural products are examples of commodities. But what exactly is a commodity and what are its characteristics? A commodity is an undifferentiated product—a good that can be interchanged with other goods of the same type. Therefore, in most cases, the buyer of the good doesn't care where the product comes from or by whom it was produced. For instance, soy from Brazil doesn't differ much from soy grown in India. And for most people eggs are just eggs, milk is just milk, corn is just corn. The same is true for palm oil, cotton, cocoa, coffee, and many other agricultural commodities. Also, consumers don't care that much whether the cotton in their jeans comes from Mali, India, Turkey, or Uzbekistan, nor do they care if the name of their coffee grower was Juan, Carlos, or Tien. What is important for most consumers is that the product tastes, wears, or looks good, and that it gives consumers what they expect for the right price.

Unlike the tourist shops, however, farmers can't rely on many different buyers that come by every day. They depend on only a few buyers and middlemen. In addition, agricultural products are usually perishable and have a much shorter shelf life than souvenirs. What is more, because of the natural cycle in which farming takes place, farmers are usually harvesting the same products at the same time (when the crop is ready). And, since you cannot always store these products because of their perishability or lack of infrastructure, everyone must sell at the same time. This causes prices to decline sharply, and every farmer loses.

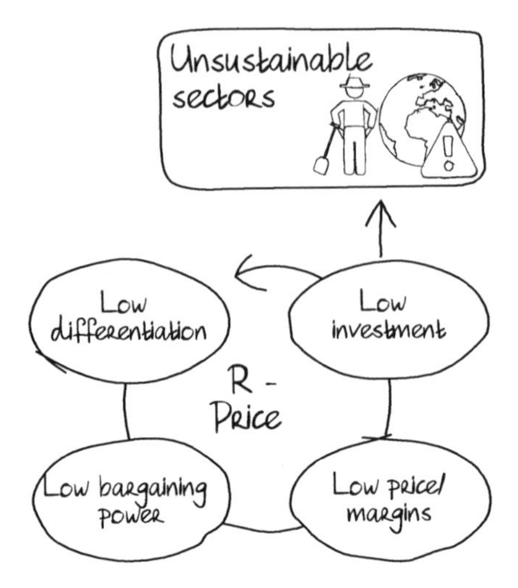

FIGURE 5 The reinforcing negative price loop of commodities

What we see in the marketplace is that most branded consumer products, such as coffee and cocoa products or jeans, are highly differentiated from each other through marketing and brands like Levi's, Nike, Folgers, Nescafé, Mars chocolate bar or Toblerone. Hundreds of millions of dollars are spent annually to convince you as a consumer that this brand is special, it is unique, you should buy this particular brand, and it is worth a higher price. But in most of the cases, the raw materials which make up the product are almost always "simple" commodities with virtually no differentiation possibilities. It is this principle that is at the heart of the first reinforcing, cause-and-effect loop, which ultimately leads to failing markets (see Fig. 5). Farmers selling simple commodities cannot differentiate in the market, so they have low bargaining power. This in turn leads to low margins and to a situation where they cannot invest much in their business.

This cycle not only affects farmers in Africa, South-East Asia, or Latin America. Farmers in Western economies are also at the receiving end of this negative price dynamic. Despite their relatively high output and exports, farmers in the West almost always complain about their low income. In the UK, farmers expressed their frustration by blocking roads. In Brussels, angry farmers dumped vegetables and tomatoes in front of the parliament. And, in the Netherlands, one of the most efficient and largest agricultural exporters in the world, enraged farmers spilled truckloads of milk onto the streets of The Hague, where the Dutch government resides.[105]

This situation of negative price spirals is also happening outside agriculture. Take the labor market as an example. For any type of work that doesn't require special skills, training, diplomas, or talent, the workers themselves are undifferentiated, and the law of supply and demand will basically determine their salary. Most of them will have little choice but to accept the wages offered to them. If they try to change this, they can be easily replaced by other people willing to do the work for the little money on the table.

Supporters of the free market system argue that this is exactly how it should be. After all, such unregulated markets are believed to stimulate and reward innovation, entrepreneurship, and efficiency, which ultimately benefits consumers. This is essentially true. However, markets that focus only on low costs in regions with failing or ineffective governments do not lead to innovation or greater efficiency; rather, they eventually lead to a massive externalization of costs and implications for others, resulting in a race to the bottom and the unsustainable economics of poverty. We should not mistake unsustainability for efficiency.

The poor bargaining position held by farmers trapped in this loop is also affected by the fact that the majority of farmers are smallholders. This means they own a small plot of land, typically smaller than two hectares. Approximately 70% of the world's food is produced by 500 million of these smallholder farmers. According to a Fairtrade report from 2013, the great majority (80%) of the world's farms are still two hectares or less in size.[106] These smallholder farmers are often unorganized, illiterate, or uneducated, and are dependent on their crops for their food supply or for cash (known as a cash crop).

In contrast, the buyers of export crops are international processors and traders, and are often among the largest companies in the world. For instance, here are some of the biggest names in the trade that negotiate prices directly or indirectly with farmers (usually smallholders). The Neumann Kaffee Gruppe is the largest coffee trader in the world. It has 2,200 highly skilled employees and a turnover of $3.1 billion (2012).[107] Or consider the Noble Group, a transporter and processor of energy, minerals, ores, and agricultural products including soft commodities, grains, and oilseeds. Its revenues in 2012 were $94 billion.[108] Another example is Cargill, which had 142,000 employees and a turnover of nearly $134 billion in 2012.[109] Bunge has around 35,000 employees, and had a turnover of $13 billion in 2012.[110,111] Bunge focuses on oilseeds and grains, produces sugar and ethanol, mills wheat and corn to make ingredients used by food companies, and sells fertilizer in North and South America.[112] ADM is another trading and processing giant, employing some 30,000 people, with a turnover of $90.6 billion in 2012.[113,114] It is active in oilseeds, corn processing, agricultural services, storage and transportation, wheat milling, cocoa processing, and food ingredients.[115] According to *The Guardian*, the A-B-C-D group (ADM, Bunge, Cargill and [Louis] Dreyfus) accounts for 75–90% of the global grain trade.[116]

Besides differences in size and economic leverage, further benefiting the bargaining position of the value chain in comparison with the bargaining position of the farmers, is the fact that many of the end buyers (the brands and manufacturers) do not require specific origins (raw material from a specific country or region). For example, buying cocoa or soy is not the same as buying an *appellation d'origine controlée* wine or champagne. Buyers are, most of the time, not that interested in where the commodity actually comes from or who grew it. Many food manufacturers have developed for their brands what is called flexible taste profiles. This is when the end product that is sold to consumers can be made using a wide variety of ingredients from various regions and origins. For example, most coffee products are constantly changing blends from different parts of the world, without consumers tasting the difference. The same is true for other products such as chocolate, cotton, soy, spices, palm oil, and almost all other commodities. The benefit for buyers is that they do not depend on a single variety or origin, allowing them to have alternatives when crops fail; but it also enables them to continuously look for, blend, sell, re-blend, re-sell, and deliver the required qualities and quantities for the lowest price available anywhere. As a result they have leverage over farmers, who often don't have this luxury.

The game of "who has bargaining power over whom" is played throughout the value chain, not just on farmer level. The traders and processors, in turn, are also under constant threat of becoming "commoditized" by their downstream clients. The large food manufacturers and brands are wary of becoming too dependent on individual traders, which could lead to higher prices. The flexible buying concept not only affects the farmers but also the traders, who can be easily replaced if prices are getting too high. And so the game of standardizing, commoditizing and gaining bargaining power over others rolls on.

An obvious effect of this flexible buying and blending is that traceability (the ability to know where the product came from) becomes very difficult or, in many cases, impossible. As a result, many brands have no real idea where their products come from. And since most consumers don't care either, the commoditization and flexible buying concept offers the most cost-efficient options for the end of the value chain.

In other cases, consumers do not know what they are consuming because the marketeers of the brands do everything possible to hide the ingredients from sight. Palm oil is a good example of such a "hidden ingredient." It is one of the world's leading agricultural commodities. Chocolate, biscuits, and cereals almost always contain palm oil. It is in over 60% of the food products we buy every day. Labeling laws have helped to hide this ingredient. Companies were not required by EU law to label products containing palm oil until December 2014. Therefore, many consumers have remained unaware that they are heavy users of palm oil.

Another example is sugar. Sugar is in almost every product we buy. Not just in sweets and cookies, but also in bread, readymade meals, pastas, spreads, and many more. But the average consumer is hardly aware of this. Many brands will do their utmost to avoid mentioning that products contain sugar, or they will try to downplay it by placing sugar lower on the ingredient list to avoid attention. They also use different names for it such as syrup, nectar, cane juice, crystals, or glucose.

According to Nick Goodall, former CEO of Bonsucro (the initiative for responsible sugar):

> Sugar is remarkably invisible for something that is consumed in such large volumes, but it is everywhere. People have a mixed relationship with that. There is an incredible amount of sugar in a slice of bread, for instance. The question is: How helpful is it to make the public aware of this ... both the bread making company as well as the consumer probably don't want to deal with the fact that there is sugar in that bread.[117]

Or take soy and corn. These are not hidden ingredients as such, but few consumers realize they are heavy users of them in an indirect way, as soy and corn are the main ingredients for much animal feed for pork, cattle, poultry, and even fish farming.[118] So if you are a heavy consumer of dairy products or meat, you are indirectly a heavy consumer of soy and corn, with all their damaging effects such as deforestation, chemical use, and bad working conditions (particularly with regard to the production of soy in South America). But most consumers have no idea.

In short, commodity markets are classic examples of markets designed and optimized to offer downstream companies (i.e., companies facing and servicing the consumers) maximum profitability. The situation worsens the further you go upstream in the value chain and where actors are more easily replaceable and margins become smaller and smaller. This is particularly the case for farmers who often live on small plots of land and have nowhere to go, are often unorganized and illiterate, and who sell undifferentiated crops to large and powerful processors and traders who, in turn, sell them to large retail and food manufacturing powerhouses. All of these forces add up to worsen an already difficult price-taking situation for the farmers (see Fig. 6).

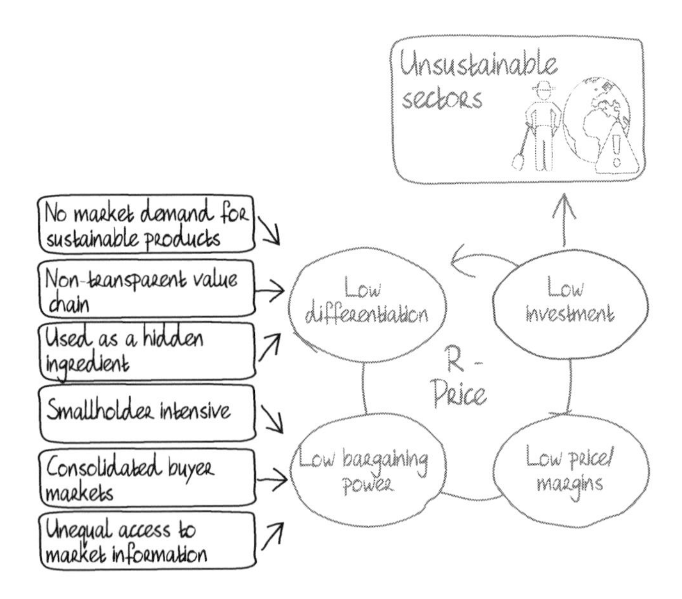

FIGURE 6 The reinforcing loops that further strengthen the negative price spirals many farmers find themselves in

Just like the buyer in the souvenir shops, the dominant behavior of buyers in the commodity markets is to seek the lowest price possible. This behavior is perfectly rational and understandable from the viewpoint of every individual trader, manufacturer, and brand, because they want to maximize their profits. However, the collective result of all individual, self-serving behavior leads to overexploitation of the farmer base, and it is starting to hurt. In particular in Africa, Latin America, and South-East Asia, this has resulted in hundreds of millions of farmers being pushed into poverty. As a result, for many crops, productivity is falling, farmers and their plantations are aging, and poverty is widespread. Furthermore, for the younger generation, almost anything offers a better prospect than becoming a farmer and living a life like their parents. This obviously threatens the future of such sectors, and many senior officers in a growing number of food companies are seriously concerned about this. But what can be done about it? This is how the market works!

4.2 The role of governments

Maybe governments are not the main driver, but they are certainly a key player in promoting and ensuring sustainable commodity production in developing countries.[119] (Hammad Naqi Khan, WWF)

As stated above, not only are farmers in the emerging or developing economies hard hit by negative price spirals, farmers in the West also feel the consequences every day. On average, however, farmers in Europe and in the U.S. are in much better shape than their counterparts in the emerging markets. This is because they receive enormous amounts of support from their governments.

For governments in the developed economies, food independence or food security is of strategic importance. According to the FAO, food security is defined as a situation that exists when all people, at all times, have physical, social, and economic access to sufficient, safe, and nutritious food which meets their dietary needs and food preferences for an active and healthy life.[120] It also means that, when push comes to shove, nations can rely on their own food production capabilities. In developed Western economies governments understand very well the devastating effects that negative price spirals can have on their farmers and food production systems. Therefore, to maintain a self-sufficient food-producing system, governments in developed economies intervene and offer a range of supportive measures to help their farmers survive and invest. You could even say that these food-producing systems are floating on subsidies. The name for this in the world of academia and policy-makers is "cheap food policy."[121] As *The Economist* observed, this "involves the government overtly pursuing policies that hold down the price of food deliberately below the competitive equilibrium price,"[122] with the objective to pursue food security, social stability, and keep cumulative price inflation of food products down.

However, the costs and the effects of maintaining cheap food and an adequate, reliable food producing system through subsidies are high and increasingly unwelcome.

4.2.1 Food shortages in Europe: never again!

Traumatized by the severe food shortages during and immediately after World War II, the first large agricultural support program was introduced in Europe in 1962. It was called the Common Agricultural Policy (CAP) and was designed to deliver on two important objectives: improve farm incomes and ensure an adequate supply of food in Europe through various direct and indirect price support mechanisms for the farmers.[123] By the 1970s, the CAP had become notorious for the resulting massive surpluses of food commodities—especially beef, butter, and wine. These surpluses could not be sold on the world market because they were so expensive. The European farmers regarded the price support of course as highly satisfactory. But the costs were and continue to be huge. European consumers had to pay well above market price for their food, while producers in the developing world found themselves unable to gain access to the highly protected EU market and, moreover, were faced in their own countries with EU surpluses dumped in their markets at artificially low prices.[124]

To this very day the CAP is still the biggest item on the EU budget. In 2011, it accounted for 40% of the total EU budget. Although the level of farm subsidies has fallen since the 1950s, in 2010 the CAP budget amounted to €43.8 billion. Direct farm subsidies are by far the largest expenditure of the CAP. Studies show that many of the recipients of CAP subsidies in Europe are not small farmers, but large landowners and agribusiness, over 1,000 of whom have become "farm subsidy millionaires."[125] Some observers joke about the costs of this policy: "If you sent every single cow in Europe on a first-class around-the-world air trip, it would still cost less than subsidizing the cattle under the CAP."[126]

4.2.2 Protecting EU farmers

Direct subsidies alone are not enough; EU farmers are also protected against competitors from low-cost countries. On average, the EU imposes agricultural import tariffs between 18% and 28% on similar products coming from outside of the EU. Compared to the EU's import tariffs on manufactured goods, which average around 3%, this is extremely high. It is even higher than the protectionist measures taken by the U.S. or Canada. These tariffs are particularly high on products that we also produce here, such as sugar, dairy products, or beef. Other commodity imports, such as cocoa, coffee, and oilseeds, are subjected to very low tariffs as they do not compete with our farmers. As a result, farmers from developing countries are seldom able to export their products to the EU unless they have preferential tariff concessions.[127] This leads to the excessive situation whereby some developing countries seeking to export beef to Europe face tariffs of up to 150%. And fruit and nut exporters to the U.S. face tariffs of, in some cases, 200% or more.[128]

The EU subsidy champions, however, are the dairy subsidy schemes. These are perhaps the most impressive, and disruptive, of all.[129]

- Dairy farmers in the EU benefit from subsidies based on a per-hectare scheme of €5 billion per year. This figure has increased from about €2.75 billion in 2005 and €4.5 billion in 2007
- EU dairy farmers are also protected by high EU import tariffs, which effectively close the EU market to dairy imports from third countries (apart from the limited volumes which enter under quota arrangements and preferential agreements)
- The EU also maintains a policy of direct intervention to buy farmers' outputs in a certain period of the year to support market prices

- In addition, in recent years the EU has initiated major "safety net" support programs for dairy farmers to sustain milk production in the face of price declines. In 2009, for example, the EU spent an additional €600 million on top of the €5 billion in direct per-hectare payments in response to low prices at the time
- The EU also pays farmers an export subsidy (or "refund") at times when Europe's dairy prices are higher than world prices to enable them to access world markets. Between 1996 and 2006, EU export subsidies on dairy products were high, ranging from €475 million to €1.8 billion
- In recent years these dairy export subsidies have been reduced and, since the end of 2009, have been set at zero. At the 2005 World Trade Organization (WTO) negotiations, it was agreed that all export subsidies should end by 2013, provided that a full multilateral trade agreement had been reached; as at mid-2014, these negotiations were still ongoing

On top of these existing measures, in 2009, the European Commission announced plans to artificially boost dairy prices by buying up 139,000 tons of dairy products (30,000 tons of butter and 109,000 tons of skimmed milk powder) at a cost to the public of approximately €300 million. In this program, the EU paid far above market prices to support the dairy industry.[130]

The inevitable effect of all the subsidies for EU farmers is artificially high production. This oversupply is often dumped outside the EU, which hits non-EU farmers hard. For example, each ton of wheat and sugar from the UK is sold on international markets at an average price of 40% to 60% below the costs of production. Every wheat farmer in the EU currently receives a subsidy of approximately €45 per ton. However, we do not allow others to play the same game. In Pakistan, for example, subsidies intended to support small-scale wheat producers have been slashed under pressure from international institutions for being unfair competition.[131] We clearly apply shameless double standards here.

4.2.3 Subsidizing and protecting U.S. farmers

The U.S. is also known for its support and protective measures to its agricultural sector. It imposes relatively modest tariffs and quotas on imported products compared to the EU, but these are impressive nevertheless. According to the WTO, the average import tariff in the U.S. for agricultural products is 8.9%.[132] In addition, the U.S. has the export enhancement program (EEP), which subsidizes exports; and then there is the Farm Bill. The Farm Bill is the big one. The Farm Bill is a repeating five-year policy which has the main objective of providing price and income support to U.S. agricultural producers. It also authorizes programs for conservation, rural development, nutrition (domestic food assistance), trade, and food aid, which are administered by the U.S. Department of Agriculture (USDA).[133] The Farm Bill was first passed in 1973 and is the major source of farm subsidies. The 2008 Farm Bill, for example, is estimated to cost more than $604 billion over ten years.[134] Its successor, the 2013 Farm Bill, was increased to $955 billion over ten years.[135] Corn, cotton, and sugar benefit the most from the subsidy programs in the U.S.

The cumulative effects of agricultural support policies in all Organization for Economic Co-operation and Development (OECD) countries cost consumers and taxpayers over $300 billion every year! This means that farmers in Western economies receive more than one-third of their income from government programs. The value of total agricultural support in OECD countries is more than five times higher than total spending on overseas development assistance (development aid), and twice the value of all agricultural exports from developing countries.[136]

4.3 What about the African farmers?

While Western governments seem to go out of their way to support and subsidize their agricultural sectors, in developing countries it is often the other way around. For most governments in the global South, agriculture is not a sector to support, but a sector to exploit. No subsidies or investments are offered, but most often taxation. As Frank Mechielsen, policy advisor at Oxfam Novib, explains:

> In the West we, as consumers, actually pay too little for our food in the supermarkets. We pay our farmers indirectly with agricultural subsidies through our tax money. In some developing countries this is completely the opposite. Over there, farmers receive low prices. Often farmers are not supported by the government. The taxes levied by the government are not reinvested in the sector. On the contrary, it is being drained from the system.[137]

Much of the relatively low involvement of governments in developing countries in the agricultural sector can be traced back to the early 1980s, the time of neoliberalism when it was believed that free markets could solve issues much more efficiently and effectively than governments. At that time, developing countries were no longer able to repay their loans to Western commercial banks. This debt crisis gave international institutions, such as the International Monetary Fund (IMF) and the World Bank, the opportunity to restructure the economies of indebted countries through structural adjustment programs (SAPs). In return for new loans, which could service their debts, SAPs enforced measures that removed barriers which, according to the IMF, hindered a freely operating, market-led economy in all sectors, including in agriculture.[138] Southern countries were forced to implement policies that included the privatization of state-owned enterprises, cuts in government expenditures, support and research programs, and the liberalization of capital markets and trade. More and more areas of public policy became subject to SAPs. The effects were devastating.

For instance, the implementation of SAPs in Zimbabwe enforced a reduction of direct state involvement in the production, distribution, and marketing of agricultural inputs and commodities, the removal of subsidies on agricultural inputs and credit, a liberalization of export and import trade, and the privatization of agricultural marketing.[139] By reforming the Zimbabwean agricultural sector, the World Bank intended to increase agricultural productivity and production, boost agricultural exports as a basis for national economic growth, and improve farmers' incomes while, at the same time, ensuring food security. However, the results turned out to be the exact opposite of what was intended.[140]

- Trade liberalization resulted in a shift to more tradable products, such as horticultural products. Eventually, this led to a shortfall in the production of corn, resulting in a lower level of food security
- Subsidies on fertilizers and other agrochemicals were removed. As prices of fertilizer rose by over 300%, the amount of cultivated land fell drastically. This led to lower levels of production
- Competitiveness of farmers was reduced by government budget cuts on infrastructure and supply chain facilities and the loss of information and statistics, which were previously provided by the marketing board
- Currency devaluation intended to foster exports led to a significant decline in the number of cattle, as imported feed became too expensive

Another example of a country affected by SAPs was Mexico. The agricultural reforms in Mexico included constitutional amendments facilitating the privatization and concentration of land and natural resources, the reduction of state participation in agricultural production, the privatization of the production and distribution of agricultural inputs and services, and the liberalization of trade in agricultural commodities.[141] Instead of boosting productivity and increasing food security, these measures eventually resulted in considerably worse conditions for smallholder farmers in rural areas, a better position for large-scale producers, less food security (food dependence went from 18% in the early 1980s to 43% at the end of the 1990s), and a significant loss of biodiversity.[142]

While international institutions forced reforms and market liberalization on developing countries, we in the West continued and even increased our subsidies and support to our own agricultural sectors. As a result, the developed economies could continue enjoying cheap resources and raw materials, while we continued to protect their own sectors from it.

4.4 No social safety nets

If you are a farmer, or a worker on a farm in developing countries, you are very much on your own. Local governments in the emerging economies are often ineffective or fail to implement effective social or environmental laws to protect workers' rights, ensure that minimum wages are being paid, ensure healthy working conditions, or protect natural resources. Even when such policies exist, they often only exist on paper and lack proper implementation and enforcement.

Fighting for workers' rights and social welfare, and environmental protection, is not a matter for governments alone. In many developed economies and democracies, public concerns are also raised and protected by local NGOs, foundations, or labor unions. However, in many of the emerging economies we see the lack of a functioning legal system that empowers and legitimizes local trade unions or NGOs to act.[143]

The absence of safety nets, standards, or effective legal enforcement, in combination with continuous price pressure from the markets, means that there are no minimum standards. These macro forces combined become self-enforcing loops which feed on and strengthen each other, creating a race to the bottom where the most unsustainable farmer wins. The use of child labor or slavery becomes a way to reduce costs and acquire more business. Burning down tropical forests for cheap expansion ensures finance and clients. The wrong behavior is rewarded and, since governments also benefit from these practices in the form of taxes or bribes, farmers have little choice but to accept the lower prices, forcing others to do the same.

Figure 7 visualizes how ineffective, or in some cases failing, governments and their policies contribute to the other feedback loops and how both loops reinforce each other.

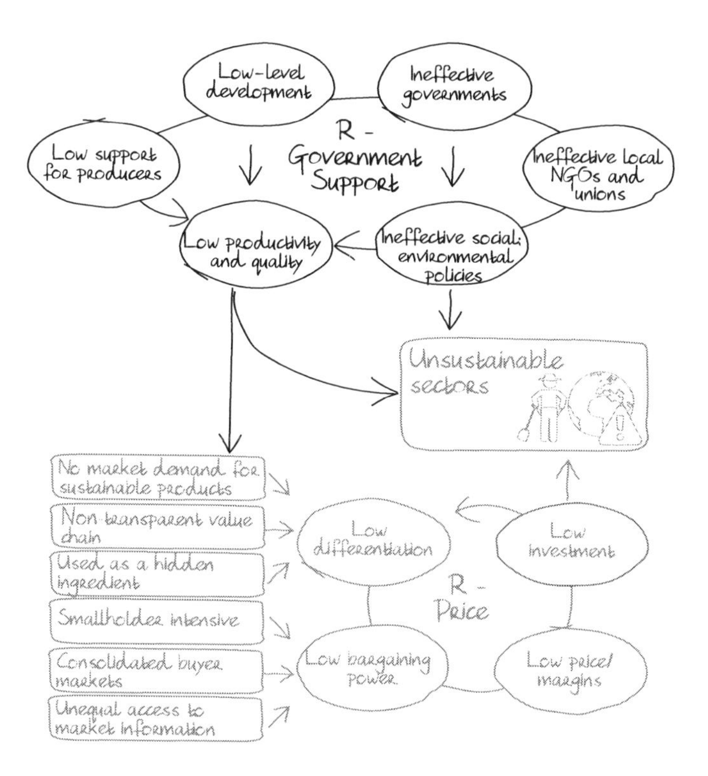

FIGURE 7 The reinforcing negative loop of failing authorities/governments

4.5 A complex chicken-and-egg situation

How do you break out of self-enforcing loops where markets go for the lowest price and governments are ineffective in giving the appropriate support and enforcing standards in the market, while at the same time Western governments are subsidizing their sectors and dumping subsidized products on emerging markets?

Many would say we need to help farmers become more professional, efficient, well-managed, and productive farmer organizations that are able to ensure a livable income for their members and members' families. This is indeed a necessary part of the solution.

However, well-managed farms need educated, trained, and entrepreneurial farmers. In many cases, the education, training, or support systems are either not there or are not as effective as they need to be. And, like any other business, well-managed farmer organizations need access to inputs (seeds, fertilizer, certain chemicals, machines), access to services (agronomical knowledge, market information, consultancy services, business services), access to finance (working capital, loans, equity, capital expenditure to buy assets), and access to markets. This is where we get stuck.

As you can imagine, not many banks, consultancy firms, or companies selling seed, fertilizer, or equipment want to do business with poor, illiterate farmers who do not make money and have no collateral. Add to that the fact that governments are also not investing in the sector, are in some cases corrupt or unreliable, and are unable to provide the basic infrastructure, and you have a perfect storm.

Would you consider doing business with these farmers? Even when farmers gain access to inputs and finance, they often lack the capacity, education, and skills to use them effectively. An important report by the International Assessment of Agricultural Knowledge, Science and Technology for Development (IAASTD) called *Agriculture at a Crossroads* presented the following conclusion: if we do find solutions to bring the right inputs and techniques to the smallholder farmers and help them to professionalize and organize, but in fact only continue to serve the already large farmers instead, then the gap between rich and the poor will continue to get bigger and bigger.[144] This is a serious problem. If it is not addressed, the situation will continue to worsen, as it further feeds and strengthens the root causes of the problem, with farmers becoming more desperate and hence unsustainable and driving down the price even further.

Some companies, suppliers, banks, and insurance companies are trying to change this, and there are examples of admirable initiatives where such parties, despite the lack of a business case, find solutions and, in some cases, even make a handsome profit. No doubt readers of this book will know of many good causes and NGOs who are really making a difference. Unfortunately, these are exceptions to the rule. The point here is that, on the whole, a poor sector makes a poor business case, and when even the basic conditions for change are absent, it makes any structural solution very difficult.

Figure 8 shows what happens when the third reinforcing loop is introduced—that of failing conditions for change.

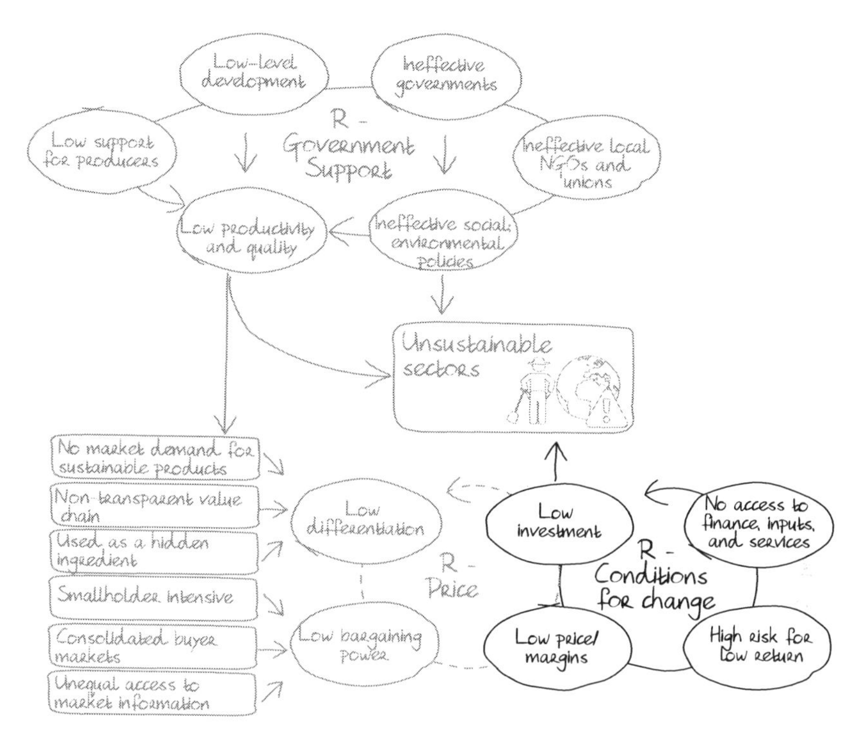

FIGURE 8 Reinforcing loop of failing enabling conditions

4.6 Unsustainability as the natural outcome

Stuck in systems like this, farmers are under enormous pressure to maintain or increase income. This is, in many cases, a matter of survival. The best way to deal with this would be by professionalizing and becoming more efficient and productive as this would be a balancing force (B+) in the model and reduce some of the negative reinforcing loops we have described. However, as we have seen, this requires education, entrepreneurship, access to inputs, knowledge and finance, and a supportive government. Since most of this is lacking, the default survival strategy of most farmers is yet another negative reinforcing loop (R-) which is to further externalize their costs as well. This comes in different shapes, forms, and sizes, depending on the sector and the context. In some cases it means clearing tropical forests so the farmer can expand cheaply and easily. Using child labor or slavery is another tactic to reduce operational costs. The same applies to not offering contracts to workers and paying very low wages. Or simply compete with your "neighbors" on poverty and accept almost any price that is offered to you under the threat that the buyer will take his business elsewhere. The implications of this externalization are felt the hardest in countries with a failing or ineffective government support system, and where farmers, with no real alternatives, are trapped in the situation. This is the perfect environment to preserve unsustainable economies.

Figure 9 shows how a failing system happens when different negative reinforcing loops are resonating together. This is the situation in which most commodity markets are trapped, and which explains the global challenges we are facing.

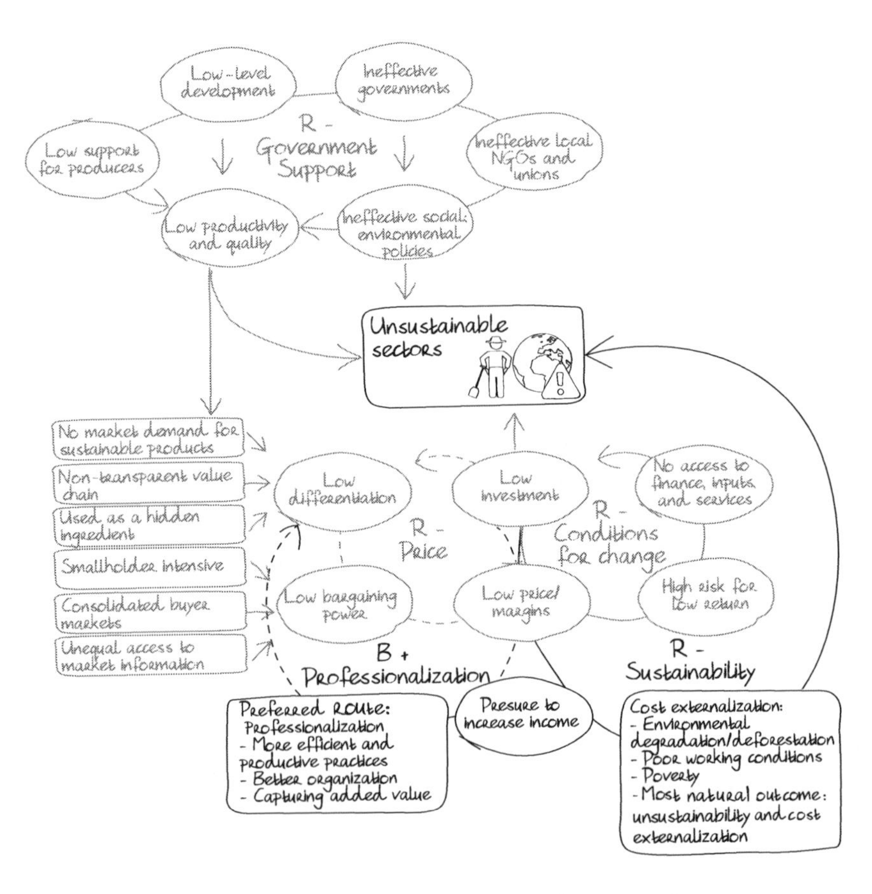

FIGURE 9 When markets fail unsustainability is the natural outcome

4.7 A cynical conclusion: we benefit from unsustainable economies

The cynical conclusion of this analysis is that when most of the agricultural sectors compete on externalizing costs, you could argue that the price of the commodities is artificially too low. For example, the prices of cocoa, coffee, cotton, soy, or cut flowers would be substantially higher if farmers earned a decent income and did not live in poverty, if children went to school, if working conditions were healthy and safe or if the environment was protected. The price of these goods is artificially low because all these things are not included in the cost price.

Because prices are artificially low, using simple supply–demand logic, it can be argued that demand in Western markets is artificially too high. From that perspective you could easily argue that we have created industries that flourish and depend for their success and growth on maintaining a situation of unsustainability, externalizing costs and respective artificially low prices. In a way, we in the West are dependent on this situation to continue, otherwise the price of food and raw materials would be significantly higher and profits and employment would drop. But also export and import taxes would drop, hurting government budgets as well. And thus we need to maintain this situation.

The analysis becomes even more cynical when public money from Western economies is used in the form of subsidies, donations, aid, or other co-investments under the banner of sustainability and development aid. This money is used to "help, support, and train" poor farmers and protect the environment. However, because it patches up the worst effects of the failing system, it also prolongs this artificial situation, benefiting the industries and the governments again. At the same time, we ease our conscience by claiming we are doing something good.

If the true costs of a sustainable sector are taken into account, prices would rise and, as a result, the global markets for coffee, spices, chocolate, meat, flowers, and many other commodities would probably be smaller than they are now. Thus, many everyday products that we are accustomed to being readily and cheaply available would become more akin to luxury goods. Perhaps that would not be a bad thing considering the exploding levels of obesity we see in the West as a result of the cheap food policy and mind-set.

The fundamental dilemma in many markets nowadays is that, on the one hand, these industries depend on low commodity prices (which are the result of externalized costs), but, on the other hand, this makes the sector unsustainable (not future-proof) and unstable. Taking into account the poverty levels, the loss of biodiversity, levels of deforestation, and loss of topsoil, and the fact that young people don't want to be farmers any more, it becomes clear we have reached the end of this exploitative model. Something has to structurally change, and this means something has to give.

The solution, of course, is that markets should start to appreciate and reward quality and the price offered for products should reflect the true costs of production, including the costs of respecting the basic human rights of farmers and their workers, and protecting the environment. These extra costs should not come in the form of development aid or in support to patch up the worst externalization effects; instead, they should be included in the price of the product. The quality product should be produced by professional farmers who are able to run their farms as businesses and make an income from them that allows them to take care of their farms, fields, and families. Of course, national governments should see the importance of their agricultural sector, take their role seriously, support their own sectors, and enforce minimum quality and living standards. Only then we are on our way towards truly sustainable agricultural economies.

But this would require a fundamental change in the way we operate now. In Part II of the book we will describe what this kind of change process looks like.

Reality is even more complex

I have limited the analysis in this chapter to the basics of the most powerful and dominant feedback loops in global agricultural systems, which are already complex. However, the reality is even more complex. The story could have been extended and completed in many ways: for instance, by describing the role of the World Trade Organization (WTO) in more detail. In its drive to promote free trade, the WTO hampers the acceptance of basic global sustainability standards and norms, which it sees as free market distortion. Another example is the large (and often bureaucratic) programs of the World Bank Group and other multinational institutions such as the FAO, International Fund for Agricultural Development (IFAD), and UN, often with national governments that in many cases seem to be used as political instruments. Enormous sums of money are spent on these programs and provide government-to-government support, but they often leave the failing market mechanisms intact; as a result, the impact of many these programs is questionable at best and leads to dependencies.

The story could also have been expanded to include the effects of having a completely fragmented and competing landscape of NGOs, branch organizations, and lobby groups that all fight for their own interests, viewpoints, and resources. The analyses could have benefited from a description of the power of companies that sell agricultural inputs, such as Monsanto, BASF, and Syngenta, and the business models they use, such as patented, genetically modified seeds and related chemicals, and the effects this has on the environment and farmer dependencies.

Furthermore, the role of financial institutions, commodity exchange boards, and the powerful industry lobby that affects many levels of the debate did not receive all the attention it deserves. Finally, I did not discuss the political instability and civil unrest and wars that are affecting many origin countries, which complicates all interventions and can wipe out any investments or progress made in the past.

Going deeper into the analysis and describing these added affects would certainly have done more justice to the complexities of the problem in agriculture but, ultimately, I believe they are only additional examples of the same failing loops and their underlying dynamics. The main purpose of this chapter was to give clear insight into the dynamics and the loops that, when joined together, cause systemic failure.

4.8 When an entire system fails, who is to blame?

When an entire system fails, who is to blame? Can you blame the farmer for using child labor or cutting down the rainforest? What would you do if you were in his place? Do you really have a choice? It is a matter of survival. Besides, what would happen if you paid your workers minimum wage and gave them proper contracts, or if you didn't use the rainforest as a cheap source of additional arable land? Your costs would go up and you would lose your business. Then what? Everybody else does it. It's the traders' fault: they don't pay enough.

What would you do if you were a trader? Your business depends on the principle of "buy low and sell high," on high volumes and small margins. This is what the boss wants and your salary and end-of-year bonus depends on it. Of course, you see that farmers don't have an easy life, but what can you do? If you pay more, you lose your business—your clients are ruthless. Besides, everybody else does it; this is how the market works. It's the retailers' and industry's fault; they are only interested in the lowest price. They do not reward your quality and they do not give you long-term contracts. And don't forget national governments: isn't it their job to take care of their own people?

What would you do if you were a retailer? It is a high-volume and low-margin game. Higher prices in the store mean lower sales. The market is ruthless, and the company sourcing policy is very clear on this aspect—you have to buy at least four points under the average market price. And if the company's profit doesn't increase every year, stock prices will drop and you'll lose your job or your bonus. Besides, all the other retailers are doing it—this is how the business works. It is the fault of the consumer: they don't care, they go for the lowest price, and they continue to buy what we sell them.

And what do you do as a consumer? We are all consumers. Do you think you should pay for a failing system? Is this your problem to solve? Are you willing to pay more? Most people don't think so.

Welcome to the dance of destiny in global commodity markets.

Peter Erik Ywema, General Manager of the Sustainable Agricultural Initiative Platform (a platform in which the food industry works together to support the development of sustainable agriculture worldwide), put it as follows:

> Procurement chains work like this: the buyer will get his bonus when he manages to buy the same amount for slightly less money, while the seller receives a fee when he succeeds to sell the same amount for a little bit more. This system contains the wrong incentives and therefore people will not necessarily do what is good for everyone. Besides, people change positions every two or three years, so they have no interest in building long-term relationships with suppliers or customers at all.[145]

This is a very important conclusion. In a system where people are rewarded for short-term self-interest and where governments fail to protect and/or support the system, most of us will feel that we don't have much choice but to play along, even if it means the end of us. It really is a matter of "tell me how you are measured, and I tell you how you (will) behave." We are all players in the same game; we all respond to the same rules and incentives. In the case of agricultural systems, the natural outcome of this collective behavior is unsustainability to a level that almost threatens life itself. The same is true for climate change caused by the abundant use of cheap fossil energy or the global financial crises caused by perverse risk taking, bonuses, and profit making. The outcome in all situations is that eventually these systems will fail (or are already failing) and we will all suffer the consequences. Just like the farmers in the Tragedy of the Commons. Yet we decide every day that this is normal and we continue to play according to these rules.

We have imposed this definition of success and these incentive structures upon ourselves (the commons), and then behave like we don't have a choice. But this is not true: there is a choice. This dance of destiny is not god-given; it is man-made; and this means it can change. A structural solution only happens, however, if, eventually, we are able to collectively change the definition of success and change the related incentives systems, which will in turn change the behavior of all the players in the game. Since almost all players go for their own self-interest in the short term, changing behavior is very much a matter of aligning what is good for all with what is also good for them in the short term. Change the rules and the players will follow.

Sustainable market transformation is all about creating the right conditions so that the rules of the game can change. Nobody can do this alone and parties will need to come to an agreement, overcome their own short-term interests (at least somewhat and temporarily and it will be beneficial for them later), and work together. It is not easy; it requires a long-term view and systemic change agents and a change strategy—i.e., a market transformation strategy—but it can be done.

As we will see in the next chapters, this market transformation strategy moves through various phases. Each phase builds on the lesson learned and insights of the previous phase, while having its own characteristic agendas, barriers, drivers, partnerships, and therefore specific intervention strategies. These are the pressure points that, if you push them, will enable you to change a system.

PART II

What is the solution?

Introducing the practice of market transformation: a stepwise approach

5
Phases of market transformation

In Part I we have seen what happens when:

- Markets are designed in such a way that all actors are rewarded for seeking short-term benefits
- There is gap or distance between their actions and the consequences of their actions
- These markets operate in the context of failing government systems
- The basic conditions for change are not present

It is under these conditions that negative loops are able to feed off each other, creating a perfect storm; markets become caught in negative spirals that eventually cause the entire system to fail. In these conditions most of the actors will feel that they have no other option but to continue with business as usual, because trying to do the right thing means acting against your own interests. A famous Chinese proverb says: "If we don't change our direction, we're likely to end up where we're headed," and this is exactly the case when systems fail. If we don't change the rules of the game, we will reap what we have sown and the consequences will be devastating for everyone.

But how do you change systems that are caught in negative spirals? But, first, do we actually know what sustainable and balanced systems look like?

5.1 Moving toward more sustainable systems is about higher connectability

Overcoming complexity is, first and foremost, overcoming fragmentation and isolation in the system, creating transparency about everyone's role and contribution, and having actors work together. After all, it is due to isolation, fragmentation, and a lack of transparency that actors can seek and attain short-term gains and get away with it (they shift the long-term consequences of their actions to others and there is no [reputational] damage or repercussions for doing this). Fragmentation and isolation in a system also prevent actors from addressing overarching issues. Therefore, market transformation above all means collectively working towards a higher level of connectedness between actors that in turn leads to a higher level of "connectability" within the system. Connectability means that the group that works together has a better ability to deal with the overarching complexities than any individual actor has by itself. It is a form of creating systemic synergy. And this means that market transformation, in essence, is the art of influencing and creating the right circumstances and conditions so that individual actors with opposing, short-term self-interests are willing and able to work together to find shared solutions to shared problems. This is the secret, special sauce of market transformation.

Adam Kahane, author of the famous book *Solving Tough Problems*, argues as follows: "Complex and 'tough' problems normally are either not solved or solved by force. Either the people involved in a problem can't agree on what the solution is and will try to find a quick fix or the people with power impose their solution on everyone else."[146] But this is not a very effective way of dealing with them. Therefore, Kahane says: "We need an approach in which the actors who are part of the problem (read: the ones who create them and benefit from them) work together creatively to understand the situation they are all caught in and then collaboratively improve it."[147]

Systems and markets can change, but only if we are able to agree how the game should be played and agree to change the rules and incentives of the game to *reward the right behavior* and to *punish the wrong behavior* (notice again how different this is from how most systems currently work). In this way we are able to align the interests of the individuals with the interests of the common good; this is the basis of system change.

The problem is that collaborative action between opponents and rivals doesn't happen naturally. It can happen only when the need and urgency is high, in a neutral environment that enables the rival stakeholders to overcome their own short-term agendas, focus on the longer-term outcome and overarching issues, prevent free-riding, and get organized. Within that environment, the first step is to jointly define an overarching vision of the joint solution and desired end game. This vision must be as compelling as possible. The second step is that the rival actors agree on clear pathways toward that vision, and each actor has their own role and responsibilities. Finally, and this step is crucial, it is important to create an environment in which every actor is rewarded for playing that new role according to the new rules, and for contributing to the overarching objectives of the bigger vision, rather than aiming for individual, short-term, benefits. In short, we need to make sure that doing the right thing is also doing the smart thing.

This simple logic translates into three principles that define healthy and successful systems, which help us to understand how to transform a failing system into a successful one:[148]

1. A successful system is characterized by connection between the individual actors and with a higher, overarching purpose or mission. This overarching vision goes beyond the immediate self-interests of the actors. It benefits all. Connection also means that the actors are no longer anonymous: they have a relationship with each other and a high (or higher) level of mutual trust.

2. A healthy system has clearly defined roles, responsibilities, and accountabilities; all actors in the system know what is expected

from them. To transform a failing system into a successful one, strategic roadmaps (based on the shared vision) need to be worked out, which define the roles, expectations, and accountabilities for each actor. Clear expectations of behavior, responsibility, and accountability create connectability and are very important to avoid free-riding at the expense of others. In this new environment, the problem can now be managed continuously while the actors learn and adapt and progress is measured.

3. A healthy system provides incentives for the right behavior for all the actors, individually and jointly, to effectuate behavior that is in line with the shared vision and the defined roles and responsibilities. In this way, a system provides added value for all its stakeholders that are doing the right thing. The desired behavior is rewarded, and clear checks and balances need to be in place to ensure that actors do not fall back into their old free-riding behavior.

Imagine for a moment what would happen if we applied these three principles of successful systems to problems such as climate change, the financial crises, the Tragedy of the Commons, the failing agricultural food systems in general, or other societal problems. Picture what would happen if we could somehow enable and incentivize individual actors to take account of a higher objective and vision, besides their own short-term interests—if we could create an environment where success is defined as doing the right thing. Would that make a difference? Yes it would—massively.

Of course, this change is not easy: it won't happen by itself and it won't happen quickly. The sector must be reorganized; the sense of urgency and the willingness to act must be established, which requires the incentive structure to change. This takes time, and the change process will go through a sequence of phases: resistance, awareness, learning, and various levels of collaboration. The good news is: each of these phases of market transformation can be triggered and managed, because each phase has its own dynamics, agendas, drivers, barriers, friends, and opponents—you just need to be able to recognize and understand them, and to know which buttons to push.

Once you have learned to recognize these different phases, you as a change agent of market transformation can now write your own play or movie on anticipating change and rely on the rational economic behavior of the "dance of destiny" dynamics. You can use those forces to manipulate and bend the system rather than push or fight against it. In that sense, market transformation is more like jujutsu than boxing.[149] The trick is manipulating the opponent's force against him- or herself rather than confronting the opponent with one's own force.

Let's look more closely at the four phases of market transformation.

5.2 Different phases of system change

If you have ever been part of a big reorganization in a company, you probably noticed that most people don't like change. They don't like it because they prefer the known to the unknown. Also, they do not always agree with the need for change, so they resist it because they are afraid they might be worse off in the new situation.[150,151] In general, only two drivers will make people change: inspiration or desperation. In most cases, it is desperation (a crisis) that will make people consider the idea that the old way may no longer be the right way. This is also where the expression "never waste a good crisis" comes from: a crisis is often a rare window of opportunity to thaw out what has been frozen for too long.

Change is a learning process, it takes time before everyone has adapted to the new situation and has settled into their new behavioral patterns. It also takes time before everyone understands the new rules of the new situation and what the new definition of success is. Once that is settled, a successful change process is in place and people will become used to the new situation. Until the moment they have to change again, of course.

The important message is that change happens in sequential phases; each phase builds on the previous phase and adds its own characteristics, dynamics, and tactics. Change is a maturing and ripening process, where every new phase creates new awareness and new insights, which prepares the minds of the actors and the setting for the next phase. This is not only true for change in a company; it is also true for change in a market or in a system. Although you can certainly steer, accelerate, and influence a change process (and it is also possible for a change process to fail), it is hard to skip a phase and immediately jump to the solution. You have to, more or less, respect these phases just as you have to respect the phases of a young child growing into an adult.

Having worked in many different global change processes, and recognizing the negative feedback loops of failing markets, failing governments, and the absence of conditions of change, I began to see certain action–reaction patterns. A certain logical order for an effective intervention emerged from it. The market transformation process can be categorized into four distinct phases (see Fig. 10), and each phase addresses one of the negative loops described in Part I. These four phases are:

1. The awareness and project phase, which raises general awareness in the sector about the problems and elicits an initial response
2. The first mover and competition phase, which mainly addresses the market failure by creating incentives for the market to compete on doing the right thing
3. The critical mass and institutionalization phase, which addresses the lack of conditions for change and involves governments
4. The level playing field phase, which addresses the institutionalization and legalization of the new normal and new norms

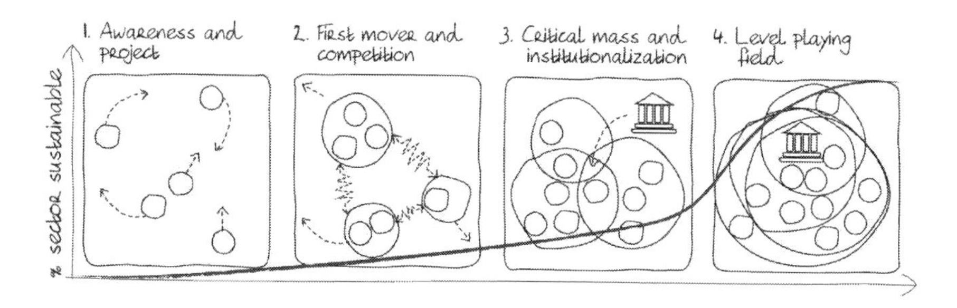

FIGURE 10 The transformation curve with four distinct phases

This transformation curve addresses the negative loops described in Part I in a sequential order starting with the market failure, and later involving governments so that jointly they can create the right enabling environment for change.

As an introduction, before we dive into some real-life examples in the next chapter, let's take a helicopter view of each of these phases.

5.3 The four phases of market transformation

5.3.1 Phase 1: the awareness and project phase

It starts with a bang. Crisis strikes; something big and ugly happens. The problem that was concealed for so long in the sector is now visible and can no longer be denied. Change agents, such as NGOs or the media, jump on the news. They will not let the organizations causing the crisis to escape this time. They analyze the situation, find the guilty parties, and hit them where it hurts. The bigger or more famous the brand or company, the better. The organizations that are under attack react as expected. They are baffled. Only a few companies will see the opportunity. Most of them deny the problem or act defensively. This is not their fault. They did not cause this. Their involvement and impact is limited. Besides, everybody is doing it, so why should they be singled out?

In an attempt to secure their future business, or protect their brand value and reputation, some companies try to make a good thing out of a bad situation. This is the phase in which companies start taking symbolic action and initiate projects. They do this partly because it is the easiest thing to do and partly because they do not know what else to do. And it works ... for a while. Money starts flowing into mediagenic and iconic projects. Compelling pictures and stories appear on company websites and in company reports. Even some of the former opponents, the campaigning NGOs, start getting involved and joint projects are launched. A project industry is born. A few years later there are many fragmented foundations and projects, each with its own focus and methodology, each claiming success and impact. However, the problem is not solved, just covered up, as fragmented projects are never the answer. It is only a matter of time before another crisis strikes and the sector is forced to move to the next phase.

5.3.2 Phase 2: the first mover and competition phase

The problem does not go away. The projects in phase 1 addressed the symptoms instead of tackling the problem at its roots. Some NGOs are still campaigning against the guilty parties, and the companies are getting tired of it. At this point, first movers start to change their game. If the problem does not go away, then they decide it is better to be the first one to solve it and turn it into a competitive opportunity. And, who knows? Perhaps they will even be able to seize first-mover benefits and force their competitors to take on subsequent costs and change their business model. Competition is the spark that drives the second phase of market transformation. With the projects implemented during the first phase, many new and better practices were discovered. This practical experience allows first-mover companies to gain a competitive advantage. When companies are rewarded for doing what is right, and if they are successful, then an "arms race" starts between first, second, and third movers. None of them wants to be seen as following their competitor, and this is why they differentiate. All try to find credible partners, and commit to more sustainable practices. This phase is characterized by a quick increase of activities and divergence of marketing efforts. What was once a non-issue, or a side project at most, now turns into programs affecting the core business of the companies.

Every self-respecting company now has to decide whether to join others' initiatives, start something new, or sit still and hope the storm will pass. In only a few years, several standards, programs, and market claims emerge. The media jumps all over these developments, and NGOs celebrate first-mover successes, while chasing the laggards. Fierce competition between companies drives constant improvement and even larger public commitments. Is the problem solved?

Unfortunately, it has not been solved, although a lot has been achieved. Competing companies alone can only do so much. On websites and reports, they claim to have made a real impact, but the reality on the ground is hard to change and, to be honest, is not entirely in the sphere of influence of companies. Remember that, for system failure to happen, three negative loops have to be in place: failing markets; failing governance systems; and failing support systems. Markets are not solely responsible for all of this, so neither can they solve all these failing loops. Besides, the marketing and promotion that drives the arms race is slowly getting old. Media and consumers are no longer really paying attention to yet another label, program, or commitment. And when there is no added value to justify the additional costs, the companies lose interest as well. But the problem still exists.

5.3.3 Phase 3: the critical mass and institutionalization phase

Something is changing. The industry is starting to understand that its old approach no longer works. More of the same competition will not solve the systemic failure. Competition is fine, but not on issues that ultimately affect everyone in the market in the same way. With this new insight, the central question is starting to change: from "How to avoid campaigning pressure?" to "How to compete and win on sustainability?" to "What does sustainability mean for our sector?" and "How do we organize ourselves and work together accordingly?"

A new question needs a new environment and new structure to answer it. To the untrained and unaware eye, nothing seems to change, but behind the scenes a new phase is emerging. This is the phase of non-competitive collaboration and it takes place on neutral ground. Neutral facilitators help competing and rival forces to work together and share their approaches and experiences. It feels uncomfortable in the beginning, to sit in the same room as your competitors surrounded by anti-trust lawyers. In the marketplace these big multinationals compete fiercely, but behind the scenes they recognize that the industry as a whole has a serious problem, it goes beyond competition and they have to work together to solve it. Not just with each other, but also with national governments at various levels. This is how a united industry can address the failing loop of governments and, together, they can address the creation of some of the conditions for change that have been lacking for so long.

Slowly, the interconnectedness and "connectability" in the sector increases, trust starts to grow, and an overarching vision and a compelling end goal start to emerge. This includes clear roles and responsibilities on how to get there. This is not an easy process. Especially in the beginning, it does not take much for individual actors to fall back into the role they know all too well—competition. Real change is now starting for the first time. This is the most sensitive and complex phase of all. In the earlier phases the industry was merely preparing itself for this structural change in working together.

While the leaders in the industry align and share an overarching end goal, it seems that not everybody is eager to join this tipping-point process. Some groups of stakeholders resist this phase and claim they have not been part of the negotiations and strategizing. Some companies will try to benefit from the situation by not joining and trying to gain cost advantages by continuing to act unsustainably. Others, like governments, NGOs, and standard-setting organizations (competitive or otherwise) from the previous phases may find it hard to accept that the industry is now heading toward more sustainable practices, but not necessarily with them. They may resist change and say that companies cannot be trusted without them. Some NGOs, which were once powerful change agents, might feel left out and will start campaigning again against the industry because they sense they are no longer the main driver of change and therefore they fall back into their former role. National governments might feel threatened because they are now asked to facilitate change that they always claimed to want, but never really endorsed. Even donors may feel threatened because, for a long time, their existence was justified by donating to various causes—they held the purse strings, which gave them a unique position of power. Now they are being asked to fund a single overarching strategy. The game is now really changing.

As the overarching end vision gains more and more followers, at a certain point it will start to influence or align with the law-making process at various levels—that is, when law-makers can codify the best practices and become a final follower, pushing the whole industry to this new level. Now sustainability can become the new normal, the standard, and a true qualifier in the marketplace. That is when we enter phase 4 of market transformation—the level playing field phase.[152]

5.3.4 Phase 4: the level playing field phase

We are slowly reaching a level playing field. In the previous phase a large group of companies worked together with local governments to implement and institutionalize the new standards and practices. Now a lobby campaign is starting to change legislation and make it part of the industry norms, so that even laggards have to comply. This phase can take a long time, but eventually governments will have to step up and codify what large parts of the industry have already implemented. The desired new standard has now become normal. This marks the end of the transition curve, but it is not the end of the change process. In fact, long before the market transformation process has entered the fourth and final phase, a new crisis, a new agenda, or a new disruptive innovation has emerged, and the transformation curve starts all over again, like endless waves lapping the seashore.

5.4 The patterns of market transformation are all around us

Once you understand the phases of market transformation, you will see it everywhere. Hopefully, after reading this book, you will be able to recognize the different phases and understand how these transformation processes can be influenced and accelerated over time. Furthermore, I hope you will be able to see what role you can play as a systemic leader and change agent.

The first part of this book focused on theory. In the following chapters we look at the market transformation curve in global agriculture in reality. We take a closer look at three agricultural sectors—palm oil, coffee, and cocoa, in changing order—and see how the different phases of the transformation curve have worked out for them. Each of these sectors is currently at a different phase of maturity in the transformation curve. Coffee and cocoa are now in phase 3. Palm oil is currently at the end of phase 2, but the signs are that the sector will move to the third phase soon, as the issues that plague the sector have not been solved and continue to haunt it. For each phase we will give short examples of other sectors that are also in this phase of the transformation curve.

So for coffee, cocoa, and palm oil, let's rewind the movie and follow these three sectors through the phases of the transformation curve in real life.

Important note about the following chapters:

It has to be said. Writing about market transformation and system change from a third-person view in these sectors is like trying to capture history. As we all know, no history book can do justice to all the angles, viewpoints, contributions, and opinions of every stakeholder who was involved in the actual events. Everybody tends to see the cause of events from their own point of view. As do I.

The real-life stories about sector transformation described in the following chapters are based on solid research, many interviews and my own experiences, perceptions, and interpretations of those periods.

I have been privileged to have been directly involved in many of the cases described in this book, but certainly not in all of them or for all of the time. Therefore, the descriptions of these cases tend to be skewed to my own experience and view on history. I do not claim that they are the only truth, nor do I claim that they are scientifically complete. The model of market transformation that is introduced here is not a scientific model—although the University of Utrecht has just completed a scientific study on the model and found it to be valid (the scientific article is expected to be published at the end of 2014)

The transformation curve model is meant to make sense of complex realities, to find patterns of behavior, to predict the next steps, and to suggest intervention strategies that can accelerate change. The sector cases are used as testimonials of how market transformation and systemic change moves through various phases of maturity, and how these phases can be initiated and shaped.

6
How does market transformation start?

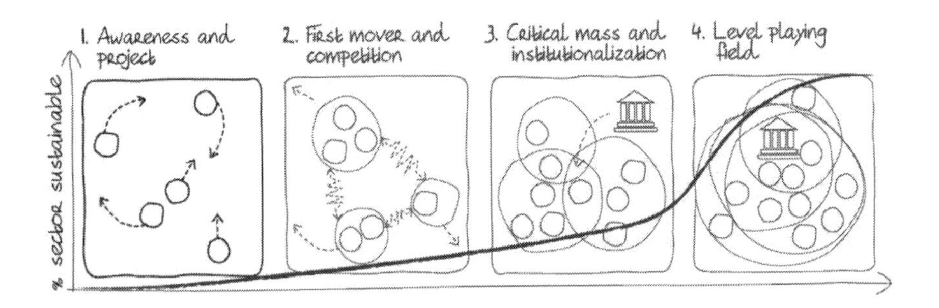

Important conclusions of this chapter:

1. Each of the sector cases has similar issues, and they are considered normal.
2. Then a crisis or an event happens. Change agents and campaigners use these events to raise awareness and build pressure for change.
3. The first reaction of the industry is often to ignore, deny, and downplay the problems, or to state that its involvement is limited and it is therefore not responsible.
4. This is followed by a period of intensively accumulating pressure through continuous campaigning and media attention.
5. To protect their brand value and reputation, some first-mover companies respond with initial projects or small symbolic initiatives, often together with other

organizations which are seen as more credible. Other companies follow and, after a while, a new industry is created where dozens, if not hundreds, of fragmented projects are implemented. All these projects claim success and impact.

6. However, the real problems have not been solved because projects can never address the root causes of the problem. Gradually, awareness dawns that more needs to be done.

6.1 Palm oil: the fire that woke up the world

Thick clouds of smoke filled the air. A suffocating heat spread rapidly. Forests, once lush and green, bustling with all types of life, burned for days. They were left smoldering until nothing but burned land and ashes were left. This was the reality in vast tracts of Indonesia back in 1997 when the country was struck by one of the most devastating series of fires in its history, affecting the entire South-East Asian region. As far away as Thailand and the Philippines, people were barely able to see because of the smoke.[153] Around 70,000 people suffered from respiratory or eye problems.

The total area in Indonesia damaged or destroyed by the 1997–98 fires was approximately 10 million hectares, an area the size of Switzerland. The massive fires caused a tremendous setback, both economically and environmentally. The regional economic costs of the fire have been estimated at $9 billion. The impact on both wildlife and several protected areas, such as national parks, has been enormous.[154] What caused this huge area to burn out of control? One explanation was that it was the year of El Niño, and the tropical forests were dry. But could this explain an area the size of Switzerland being destroyed by fire? Was it lightning that set the dry forests on fire? Was it an accident?

The U.S. Agency for International Development (USAID) assembled a team of scientists to trace back the origin of the fires.[155] With the help of satellite data and interviews with local villagers, they discovered that the fires were set deliberately by palm oil estates. This was done to clear land for rapid and cheap expansion of their palm oil plantations, and to hide timber poaching and land theft. Until 1994, the method of "controlled burning" was legal and commonly practiced in Indonesia to clear land for agriculture. However, the practice was completely unregulated. This time, the combination of illegal burning by large palm oil estates and the dry forests due to El Niño meant that the fires got entirely out of control.[156]

In the decades before the fires, Indonesia and Malaysia had been able to significantly increase their palm oil production through large programs subsidized by the World Bank. By the early 1990s, palm oil had become very profitable and no longer needed subsidized loans. Commercial banks took over and started to provide capital for expansion. Between 1995 and 1999, projects in the sector were backed by domestic and foreign investments of over $20 billion, all with the consent of the Indonesian government.[157]

6.1.1 Awareness is rising

The fires in Indonesia caused massive air pollution throughout South-East Asia region. In response, these countries, all members of the regional organization ASEAN (Association of South-East Asian Nations), developed a Regional Haze Action Plan to reduce cross-boundary pollution (see the information sheet about palm oil in Appendix 1). Faced with international pressure from other countries and the ASEAN organization, the Indonesian and Malaysian governments agreed to implement a "zero burning policy." Unfortunately, neither government effectively implemented this policy.[158]

Because the national governments in the region failed to act, NGOs took matters into their own hands. At the end of 1997, WWF issued a milestone report, *The Year the World Caught Fire*, calling for political leaders, industrialists, ecologists, and other stakeholders to work together and find solutions.[159] The message was targeted at a broad audience, including the general public and retailers, and the financiers behind the plantation expansion. Other NGOs and development agencies started to set up local initiatives and programs. One example was Project Fire Fight South-East Asia (PFFSEA). The goal of this project was to strengthen national, regional, and international networks for forest fire prevention and management. The UN Development Agency (UNDP) began working with the Indonesian government to build its capacity to assess, respond, and monitor environmental issues. And, in that same year, Sawit Watch was founded, an umbrella NGO with the mission of raising awareness about social and environmental impacts of irresponsible palm oil expansion at the local, national, and international level. Sawit Watch also supported local communities in their resistance against irresponsible palm oil production and illegal land claiming. Other international donors also became interested and funded projects to increase knowledge about more sustainable palm oil production practices. The objective of this project was to showcase that is was possible to expand oil palm plantations without deforestation. Much later, these better agricultural practices would be used to create the first standards for sustainable palm oil production.

The campaigns were effective. Global awareness about the effects of oil palm expansion in tropical forests grew rapidly. But it did not immediately lead to improved practices by the oil palm estates.

6.1.2 The blame game begins

Greenpeace, one of the more campaign-oriented NGOs, decided to follow the money trail and investigate the financial structure behind the illegal fires.[160] Jan Maarten Dros, currently International Palm Oil Program coordinator at Solidaridad, looked back on this period:

> Oil palm is a tree crop. A new oil palm estate needs about four years of growing and investment before it sees any revenue. Plantation owners therefore depend heavily on external investment. This was especially the case in the late 1990s, because there was relatively little capital in the region due to the Asian economic crisis.[161]

A simple, but effective, idea was born. If the palm oil sector was no longer able to attract investment, it would be unable to expand, and the forest fires would stop.

Greenpeace quickly discovered where the investments came from, publishing its findings in the report *Funding Forest Destruction* in 2000, which was to have a big impact. The report proved beyond any doubt that many major European, American, and Asian banks were assisting the palm oil industry. In particular, it revealed that mainly Dutch financial institutions such as Rabobank, ING, Fortis, and ABN AMRO were the second largest financiers of the palm oil industry, right after the Malaysian banks.[162] This was the smoking gun the NGOs were looking for, and the Dutch banks (which at that time were all promoting themselves as sustainable and responsible) were caught red-handed. The ideal campaign target was found—Dutch financial institutions.

The Dutch branches of Greenpeace and Friends of the Earth immediately started campaigning against the banks. They demanded that the financial institutions comply with a list of conditions for investments in oil palm plantations. The conditions included the following: no cutting and burning of forests for new plantations; respect the rights and wishes of local communities to avoid social conflicts; and no violation of Indonesian laws. The pressure on the banks was further increased by a campaign during which 250,000 postcards were distributed. The cards made it easy for account holders to demand that their banks stop investing in environmentally damaging plantation projects.[163]

It did not take long for the banks to react. And their reactions were predictable. On the one hand, they recognized the problems in the sector and agreed to inform Asian financial institutions and their clients about the issues. On the other, they avoided taking responsibility, using the argument that their individual role in financing the sector was limited and they could not be held accountable. Jan Maarten Dros described this period as follows:

> At the time, the Rabobank was advertising its sustainability profile on national television. And then they were accused of enabling the forest fires. However, their vested interest in palm oil was so big that they claimed it was impossible to disallow the slash and burn practices their clients practiced. There was supposedly no other way to conduct business. However, ABN AMRO, the other major Dutch bank, had a far smaller position in agriculture and in palm oil. They also had this high-profile Corporate Social Responsibility agenda. They must have thought: we don't need this 1% in our portfolio if it results in this kind of corporate brand risk, let's get rid of it and use this to our advantage to appear more sustainable than the Rabobank.[164]

Consequently, ABN AMRO was the first bank to sign the code, and NGOs rewarded the bank by publicly praising it for this policy change. The other Dutch banks had little choice but to follow, albeit hesitantly. In the subsequent months, Germany also implemented new investment guidelines in the forestry sector prohibiting the conversion of High Conservation Value Forest into arable land by German development projects.

6.1.3 What about the companies who buy palm oil?

In those days, surprisingly perhaps, the food industry as a buyer of palm oil was not targeted by the NGOs. This would change significantly a few years later, when big food brands would become the main target for NGO campaigns, such as the fierce attacks on Nestlé's KitKat brand and, more recently, the attack on Procter & Gamble's Head and Shoulders brand.[165,166] When the first campaigns against the banks began, the perception was that the major brands were buying palm oil from long-existing plantations, and therefore they were not the main drivers of expansion through illegal burning.

Although not a target at the time, other companies were carefully watching events in the palm oil industry. Robert Keller, Head of R&D at Migros, a major Swiss food retailer, became interested after reading an article in a Swiss newspaper entitled "Instead of Tropical Wood, Borneo Delivers Margarine."[167] Keller realized that the issues concerning palm oil production in Borneo had a major impact on Migros, since palm oil was used in so many of its products. Migros started to look for ways to reduce its dependence on palm oil by substituting sunflower oil. Later on, Migros contacted WWF Switzerland, with whom it had been working on sustainable wood purchasing since 1997. The company wanted to join forces to develop a list of criteria for sourcing sustainable palm oil which would be used in Migros factories.[168,169]

Another food giant that watched the development carefully was Unilever. Unilever is a leading consumer goods company and one of the largest buyers of palm oil in the world. It sources around 1.5 million tons of palm oil every year (around 3% of the entire world production). In 1998, a year after the forest fires, Unilever started the Sustainable Agriculture

Initiative to ensure its continued access to key agricultural raw materials. Palm oil became one of the priority products in this initiative. To improve its understanding of sustainable agricultural production, Unilever wanted to define indicators for sustainable production, which it would test in pilot projects on its palm oil plantations. In 2000, Unilever established the Sustainable Agriculture Advisory Board and invited NGOs to join, including WWF as environmental expert. This was the beginning of a strong alliance between Unilever and WWF. That same year, Unilever accepted an invitation from WWF to form a partnership with others to develop economic, environmental, and social criteria for sustainable palm oil production. This eventually became RSPO—the Roundtable on Sustainable Palm Oil (discussed in more detail in the following chapters). According to Jan Kees Vis, Global Director of Sustainable Sourcing Development at Unilever and founding Chair of the RSPO, this became "a relationship that continues till this very day."[170]

The 1997 fires in Indonesia woke up the world and sparked international outrage. For the first time attention was paid to the destructive practices of the palm oil sector. The campaigns and reports were effective in revealing to the public that oil palm estates, local governments, and banks profited heavily from the situation. The problem could no longer be ignored. It was the beginning of the awareness and project phase for the palm oil sector.

6.2 Poverty in your coffee cup

The world runs on money, oil, and ... coffee. Hardly anyone can go without this stimulating beverage; the coffee sector is enormous. Coffee is grown in about 70 countries around the world.[171] It is the most widely traded tropical agricultural commodity and, in terms of sales value, it is the largest commodity market after oil! Every day approximately 1.6 billion cups of coffee are drunk.[172] Amongst the heaviest coffee drinkers are Scandinavians and the Dutch, consuming on average between 8.5 and 12.5 kg of coffee per person per year.[173] Europe is the largest coffee market in the world, consuming 40% of total coffee production, followed by the U.S. with 16% and Japan with 5%. More than 25 million farmers and their families are directly dependent on the production and processing of coffee beans. Most of them are smallholders.[174] On the buying side, the industry is highly concentrated and is dominated by big companies such as Nestlé, Mondelēz (former Kraft Foods and recently taken over by D.E. Master Blenders), D.E. Master Blenders 1753 (former Sara Lee/Douwe Egberts), J.M. Smuckers, and Strauss Coffee. Together, they control approximately 45% of the global market. A truly global industry is behind your morning cup of coffee.

Although many people and economies depend on coffee, and much money is earned in this sector, the production of coffee is characterized by severe problems. Poverty, bad working conditions, child labor, and environmental degradation are a daily reality. The case wasn't always this bad. There have been years in which coffee actually was a relatively stable source of income for most coffee growers and their producing countries. These were the years of the International Coffee Agreement (ICA). The ICA was for the coffee sector then what OPEC is for oil-producing countries today. The signatory countries agreed how much coffee they would produce to maintain higher prices and stabilize the markets. The agreement started in 1962 and worked reasonably well until it collapsed in 1989 when ICA members failed to reach agreement on export quotas. The situation slowly went downhill from there.

The breakdown of the ICA led to a situation in which governments no longer discussed and agreed upfront to support the interests of the entire sector. Instead, each coffee-producing country opted for its own short-term interests. With the quotas removed, the producing countries all increased their production to benefit from the relatively high coffee prices. This led to a rapid increase in the coffee supply, with falling prices as a result. The situation for coffee farmers became worse and worse. Meanwhile, the industry and the importing nations benefited from the low coffee prices. Coffee was cheap and plentiful. Life was good for the downstream coffee industry, but increasingly difficult for the farmers, who became trapped in poverty.

6.2.1 Fairtrade coffee to the rescue

Not everybody was blind to the poor living conditions of the farmers. Several NGO and church-driven projects began to help smallholder coffee farmers. The most famous initiative is probably Max Havelaar, the original fair trade concept, which was launched in 1988. The initiative was started by Nico Roozen from Solidaridad and Frans van der Hoff, a priest living in Mexico. The name "Max Havelaar" was derived from a character in a famous Dutch novel who opposed the exploitation of coffee pickers in the Dutch colonies.[175] The Max Havelaar Foundation worked with a code of conduct that helped coffee farmers to organize themselves better and ensured that they would be paid a minimum price for their product, and that they would receive payment in advance. This concept went against conventional thinking and was the opposite of practice in the coffee industry at the time. The first certified coffee was sold in churches and charity shops, and later in Dutch supermarkets. Max Havelaar would later become the International Fairtrade movement.

The coffee roasters in the Netherlands, and later other countries as well, did not welcome the Max Havelaar model. They considered it strange, disruptive, and too much of a niche, do-good model. Stefanie Miltenburg, currently Director International Corporate Social Responsibility at D.E Master Blenders 1753 and Director of the Douwe Egberts Foundation, explains what Douwe Egberts thought about Max Havelaar:

> We kept Max Havelaar outside of our company. It was a matter of principle. The board of directors at the time didn't agree with the Max Havelaar minimum price model. According to them, Max Havelaar didn't respect the market mechanism of supply and demand. The NGOs kept insisting that we should buy Max Havelaar certified coffee. If we would do this, we would no longer be blamed for the bad conditions of coffee farmers, the NGOs said. But we refused. If we were going to do anything about it, it would be through a different initiative than Max Havelaar.[176]

In those days, Annemieke Wijn worked as Senior Director for Commodity Sustainability at Kraft Foods (now Mondelēz International), an international competitor to Douwe Egberts. Kraft Foods took a similar stance as Douwe Egberts. According to Wijn: "There wasn't that much of urgency, a feeling of we have to do something about this. The Max Havelaar concept was not convincing to us."[177] The idea of paying a minimum price did not seem logical to the big coffee brands. They thought it would only lead to more coffee production and therefore to even lower prices, making the situation worse. They also disagreed with the notion of buying only from smallholders and not from the larger coffee estates. Nico Roozen, co-founder of Max Havelaar, described the resistance of the companies:

> We had many disputes with the board of directors of Douwe Egberts. I felt strongly that DE had to collaborate, but they disagreed. They were convinced that their responsibility was limited to the four walls of their organization. They really believed they were not responsible for the fate of the coffee farmers. At one time, we even brought in a delegation of Mexican coffee farmers and arranged a meeting between them and the DE management. But DE was deaf to our arguments. Their knowledge of the situation in origin countries was minimal. The trade was anonymous, and many believed that paying the market price (or below) was the right thing to do.[178]

Miltenburg describes her view of the events: "At that time not many of us at Douwe Egberts, the biggest coffee roaster in the Netherlands, visited a coffee farm regularly. Our buyers bought mainly through trade partners, so we had little direct contact with the farmers."[179] And this was true for most coffee companies. Most of them had little knowledge of what was actually going on the farms.

6.2.2 Pressure increases

However, the ongoing campaigns, questions, and visits from the NGOs gradually raised the companies' awareness about the issues. Stefanie Miltenburg was asked by her CEO at the time to investigate what the main issues were in the coffee-producing countries, and what Douwe Egberts could do to take responsibility and help the farmers. In the late 1990s, she traveled to Uganda, one of their biggest suppliers of good-quality Robusta coffee. To help the coffee farmers, DE started with a typical project approach, as Miltenburg described:

> Douwe Egberts started to set up its first local cooperation projects with farmers to learn what the role of the company was and what was needed to make the necessary changes. Projects were set up with a merely philanthropic character. People were living in poor circumstances, so we decided to start with basic

> investments in the community together with an NGO. We provided them with boreholes for water and with cleaner cooking stoves. This had little to do with actual coffee production. We just wanted to help the communities.[180]

According to Annemieke Wijn, Kraft Foods responded similarly to the continued pressure at the time: "Our activities were driven in the first place by this potential public issue. We started to look at what we could do, and we started projects. We had some projects in Peru, Colombia. These were tiny projects trying to help the farmers and their communities."[181]

6.2.3 Then came the big turnaround

With balloons, music, and drinks, Douwe Egberts celebrated its 250th birthday on June 11, 2003. This was certainly an important moment in its history, and many festivities were planned. However, the party was about to be disrupted. On the day of the celebration, a Dutch newspaper ran an article with the headline "Douwe Egberts Buys From Slave Plantation."[182] The article was the result of a larger campaign that Oxfam Novib had started ten months earlier, in September 2002. Oxfam then published a report, *Mugged: Poverty in Your Coffee Cup*. The report claimed that, over the past five years, the price of coffee had fallen by almost 70%, forcing many coffee farmers out of business: "Small coffee farmers in developing countries sell their beans for less than they cost to produce. Meanwhile, the largest coffee corporations continue to reap enormous profits."[183]

Miltenburg remembered the day clearly:

> It was not very nice to be attacked on your birthday. The worst part for Douwe Egberts was that the company was unable to dispute these allegations, because in this particular case (a coffee plantation with potential slavery in Kenya) we didn't know whether we bought coffee there. There was no transparency in the supply chain.[184]

This campaign, as we will see later, would prove to be a turning point for Douwe Egberts.

For Kraft, the turnaround came when their CEO took interest in the cause, said Annemieke Wijn:

> Our CEO came down and looked at coffee growing areas himself. This was around 2003–2004. It was extremely important to have the senior leaders come to El Salvador and look at and

talk to producers. The issue suddenly got a completely different level of attention. They spent two days going around the place and came back and publicly formulated a sustainability strategy of some sort. The CEO publicly said: "Coffee sustainability is important to us."[185]

6.2.4 Now that we have got your attention

The coffee crisis had now reached the boardrooms of the big coffee brands. The NGO campaigns and the media attention were starting to hurt brand value and could no longer be ignored. Due to long periods of low prices, coffee farmers had become very poor, the sector was becoming unattractive to work in, and farmers everywhere were abandoning their farms. Coffee farmers refusing to grow coffee had not been seen before, and it scared the companies. They had grown very accustomed to cheap and abundant coffee. Clearly, helping rural communities and coffee farmers with projects was not solving the problem. What should they do?

As a result, the coffee sector could no longer deny the issues in their sector. With this awareness, and with the project phase finally started, the sector began making progress: a few years later the coffee sector would become a leader in the sustainable market transformation movement.

6.3 How it started in the cocoa sector

For many people, the taste of chocolate melting in the mouth is the ultimate indulgence. Unfortunately, cocoa does not taste that sweet on the other side of the value chain. While chocolate is seen as a luxury product in Western countries, it is treated as a common commodity in the countries producing it: the lowest price wins. Due to a number of factors, such as plant diseases, old and unproductive cocoa trees, inadequate fertilization, and traditional and outdated agricultural practices, the productivity of cocoa per hectare in West Africa is very low and, in many cases, even declining. As a result, the average income of a cocoa farmer is often well below the absolute poverty line of $1.25 per day.[186] Child and forced labor, deforestation, and severe poverty are common in many cocoa-producing areas, particularly in West Africa.

6.3.1 Diseases spread

The problems in the cocoa sector go back a long way, as do attempts to do something about it. In the late '80s and beginning of the '90s of the last century the sector was seriously affected by diseases and pests, most notably by the outbreak of Witch's Broom disease in Brazil. The Witch's Broom is a fungus disease that reduces the ability of trees to produce pods with beans. The only remedy is to kill and take out infected trees, wiping out the complete farm, a measure which resulted in economic hardship in Latin America in the 1990s. It decimated cocoa production in Brazil, which fell from its position as the world's second largest cocoa producer in 1992 to being the sixth largest producer today.

With the downfall of Brazil as one of the largest producers of cocoa, the cocoa industry became concerned about the future availability of cocoa. It realized its vulnerability and understood that when a plant disease as devastating as Witch's Broom disease can spread so fast and wipe out one of the major producing countries, it only be tackled when you join forces, work together, and take action. This eventually led to the establishment of the World Cocoa Foundation (WCF) in 2000.

In the words of Bill Guyton, President of the WCF:

> WCF was established in 2000 and at that time the major concern was that diseases and pests like the Witch's Broom fungal disease were destroying cocoa trees in Brazil. Some of the leaders in the industry, the larger companies, came together at that time to see what more could be done to support cocoa farmers. WCF was a way to think strategically as an industry about what kind of activities and interventions could be helpful to better reach cocoa farmers and make sure they were successful.[187]

This collaborative approach between competitors proved that for the industry the threat of diseases was more powerful than their competitive instincts.

One of the first actions the WCF and its members took was the start of the Sustainable Tree Crops Program (STCP). The STCP was a large public–private partnership program that provided a framework for collaboration between farmers, the global cocoa industry, the local private sector, national governments, NGOs, research institutes, and development investors. The objective was to improve productivity and growth in rural income among tree crop farmers in West/Central Africa.[188]

6.3.2 Public outrage

Attention to the poor social and working conditions of cocoa farmers also date back to around the mid or late '90s. Already by 1993, Max Havelaar had started selling fair trade chocolate in Europe, and in 1994 it entered the UK market (for a more detailed timeline, see the information sheets on cocoa in Appendix 1). However, fair trade cocoa remained a small niche market for a long time. This changed slowly after Unicef published a report revealing child labor practices on cocoa farms in Côte d'Ivoire (Ivory Coast). A short time later, in 1998, the BBC broadcast the documentary "Slavery: A Global Investigation." The public was shocked about the extreme forms of child and forced labor in Côte d'Ivoire's cocoa sector. The Côte d'Ivoire government rejected any allegations that child labor was widespread on cocoa farms and called the accusations "nonsense" and "wildly inaccurate."[189] A few months later, however, the report *A Taste of Slavery: How Your Chocolate May Be Tainted* was published by the Knight Ridder newspapers. Written by Sudarsan Raghavan and Sumana Chatterjee, it would later win the George Polk award for international reporting.[190] About the same time, the UK newspaper *The Daily Express* published an article, "Chocolate, It Seems, Carries Modern Slavery Into Our Homes."[191] The fire of public awareness had ignited and was about to intensify.

From that time on, NGOs launched numerous campaigns, some of which are still ongoing. The cocoa industry was now globally accused of profiting from child slavery and human trafficking. The European Cocoa Association, the branch organization for cocoa and chocolate in Europe, dismissed the accusations as "false and excessive," and the industry said the reports were not representative.[192] But these efforts were in vain. The mounting pressure on the chocolate sector led, in 2001, to the Harkin–Engel Protocol, an international agreement led by the U.S. aimed at ending the worst forms of child labor (according to ILO Convention 182) and forced labor (according to ILO Convention 29) in the production of cocoa, no later than 2005.[193]

With the industry under continuous NGO and media attacks on social issues, it didn't take long before the World Cocoa Foundation platform was also used to get the message out about addressing the worst forms of child labor—as Bill Guyton remembers:

> The child labor issues made companies realize that in addition to looking at the productivity (and agronomical) side of cocoa sustainability they also needed to look more broadly at the social dimension. It helped to expand the thinking of companies on how to have an approach that integrates the social and economic issues.

6.3.3 Insufficient progress

Despite the different projects and programs that were started within WCF or as standalone projects, the results lagged behind the intentions, particularly on the issues of child labor. In July 2005, the deadline passed for the Harkin–Engel Protocol, and it was time to measure impact. According to the evaluation of the protocol, the cocoa industry had made some progress and had met some of the objectives. For example, one of the objectives was that the cocoa industry would stop denying and acknowledge the problems of child and forced labor in cocoa production. And so it did. The industry founded the International Cocoa Initiative (ICI) in 2002 with the specific mission to eradicate child and forced labor on cocoa plantations, and by 2005 $3 million had been spent on pilot projects (see the information sheets on cocoa in Appendix 1). In addition, in 2004, a child labor verification working group was established. But these efforts were nothing compared to the seriousness of the problem or compared to the profits the industry makes annually.

Independent research demonstrated that children were still working in cocoa production, performing dangerous tasks, and regularly missing school. Also, the cocoa industry was criticized for failing to implement certification standards. Overall, the U.S. Congress wasn't satisfied with the cocoa industry's progress.

In 2005, it escalated. The International Labor Rights Fund (ILRF) filed a lawsuit against Nestlé, Cargill, and Archer Daniels Midland (ADM) on behalf of three Malian children. Despite the fact that none of these companies actually owned cocoa farms themselves, the ILRF accused the companies that they were (partly) responsible for the fact that these children were trafficked to Côte d'Ivoire, forced into slavery, and experienced frequent beatings on cocoa plantations. Although the claim was dismissed in September 2010 by the U.S. District Court for the Central District of California after it ruled that corporations cannot be held liable for violations of international law, it scared the cocoa industry.[194]

The Harkin–Engel Protocol and the lawsuit were real eye-openers for many companies. It was now becoming painfully clear that sustainability issues in the sector would not go away. They had to do more than set up and join umbrella organizations and spend money on pilot projects and programs. Not only the future of their sector but also the reputations of their brands and companies were now at stake. The sector had to find more credible and holistic ways to tackle the issues and to regain the trust of the public and the government.

This marked the start of a serious arms race between cocoa companies regarding efforts towards sustainability in which almost every major cocoa and chocolate company in the industry would compete on their individual sustainability programs, standards, and labels. As we will see in the next chapter, the intensity of this race was higher than in any other sector.

6.4 Reflections on the first phase of market transformation

The forest fires in Indonesia, and the campaign that followed, woke the world up to the devastating impacts of oil palm growing practices. Likewise, the coffee crisis led to campaigns that forced the coffee industry to act. Similar crises and events triggered similar reactions in the cocoa industry. If we look at these developments, a pattern seems to emerge in terms of what triggers change and how an industry reacts to accusations. This is the start of the awareness and projects phase, the first phase of market transformation.

In each of the sector cases, the unsustainable practices were not new. They had existed for a long time, but were seen as a normal part of how business was done. Then a disaster, a crisis, or an event struck the industry and this altered the status quo. An organization or an individual outside the industry stands up and begins raising awareness. These outsiders are essential to the process; they do not benefit from the old situation and are, in the eyes of the public, credible change agents. They can blow the whistle, point the finger of blame, and reveal to the public what was known to insiders all along but was not given a high priority. Industries are proven to be profiting from bad practices, and others are suffering the consequences. In the case of the Indonesian forest fires, the banks, the oil palm estates, and the Indonesian government profited directly from the way the industry was expanding, but they did not feel the consequences directly. In the case of coffee and cocoa, the value chains, brands, and local governments—through export taxes—benefited for many years from an abundant and cheap raw material. As a result, the brands could keep prices low and increase their sales, while still maintaining a very healthy profit margin. Most people turned a blind eye to the fact that their sectors were running on exceedingly cheap forced labor or child labor and that the farmers were living well below the globally agreed minimum poverty line. When companies benefit from something, they are unlikely to change it, particularly when everyone else is also doing it.

In this phase, even if companies were aware of the problems and are willing to change, they find it hard to do so. Changing unsustainable practices means higher costs or fewer sourcing options, while competitors can continue business as usual. Moreover, consumers are unlikely to reward these efforts; most will prefer to buy cheaper products from a competitor. As a result, good intentions could be punished by the market. It is much easier to continue business as usual. The sector is trapped in a situation where bad behavior is rewarded and good behavior is punished.

6.4.1 A good campaign: no pain, no gain

A well-prepared and -executed campaign can have an enormous impact. It can be an effective catalyst for change, as illustrated by Daudi Lelijveld, former Sustainability Director at Cargill and former Vice President of Barry Callebaut's Cocoa Sustainability Programs, the two largest cocoa traders and processors in the industry: "There was an awful lot of fear amongst the brand companies, in particular about being targeted by NGOs, especially from a reputational risk perspective."[195] Once a brand becomes "infected" with a negative story, that fear becomes justified and it is hard to restore public trust. Most companies will opt for "brand damage control," and either defend themselves against the accusations, or simply comply with the demands of NGOs.

A defensive response to allegations is quite common for companies or governments in this phase. Such a response is often a denial: "It's not our responsibility what happens in other countries: that's the responsibility of the local government"; or "We can't change the problem: we are only a small part of a larger chain and it is too expensive"; or "There is simply no other way to conduct business." Alternatively, companies can remain silent, hoping the issue will simply go away.

As pressure continues to increase, at some point most companies will choose to protect their brand and reputation. Uneasy with the situation, and uncertain about what to do, most companies will start a wide array of projects. These can range from research, development of codes of conduct, donations to charitable projects, setting up their own foundation, or starting to execute projects in the field themselves. In many cases, companies will choose to work together with NGOs or other "credible parties," sometimes even with the same NGOs that attacked them in the first place. These new partnerships benefit both parties. Companies learn from the NGOs' experience and they can increase the credibility of their projects. At the same time, NGOs benefit from the resources of companies, which fund their projects. Also, the NGOs can claim that they persuaded the companies to take responsibility. David has defeated Goliath.

Being the first company to accept accusations and comply with the demands of NGOs (first mover) can offer a competitive advantage. In the case of palm oil, it was a strategic decision for ABN AMRO to accept a relatively small loss and promote itself as more sustainable than the Rabobank. NGOs understand this aspect all too well and will use this first-mover advantage by promoting the heroes, using them as a role model to exert extra leverage on laggards: "They are moving, why aren't you?" This is the public crowbar that will eventually break their resistance. At some point, the other companies will also have to move. Not out of conviction, but to avoid even more damage. And most of them will seek their niche in the project landscape: they will launch more projects. And so more and more companies join the sustainability project race.

Most of the time, the projects have more promotional value than real and structural impact on the farms. In general, projects lack the scale, structure, or resources to really affect the root causes of the problem. The project may serve the immediate needs of the poor, but does not lift them out of poverty. For this, more needs to be done or, as we will see in the following chapters, it is a matter of time before the problems return.

6.5 Examples of other sectors in this phase of market transformation

Currently (2014), most agricultural sectors have moved beyond this first phase of market transformation. Not many sectors who are dealing with sustainability issues have been able to withstand the pressure and not act. An example of a sector that is currently at the end of the first phase of market transformation (or at the very beginning of the second phase) is perhaps the spices sector.

6.5.1 The spices sector

The spices sector cannot really be called one sector as there are thousands of spices in the world, some used just locally, all with their own aromas and tastes, but also with their own issues and problems. About 50 spices are of international importance, of which pepper, cinnamon, nutmeg, ginger, cloves, and turmeric are some of the biggest in terms of sales around the world. A lot of spices are now successfully grown across Africa, but many originate from China, India, Indonesia, Latin America, and the Caribbean. The sector is relatively small compared to the others, but it is still a billion-dollar industry.[196]

What are the main issues? These primarily concern aspects of economic unsustainability, such as low income, child labor, and unhealthy working conditions. In some cases, there are other issues such as loss of biodiversity and chemical use.

Where is the sector now? The sector is at the end of the first phase, or at the very beginning of the second phase of market transformation. It still has some way to go. The sector has several sustainability standards, such as Fairtrade, GlobalGap, EU Organic Labels, and Naturland, but there is little competitive dynamics on the issue of sustainability. Most companies are engaged in social projects. There are also few campaigns and little public pressure on the sector and therefore the feeling of urgency is relatively low. Many companies in this sector are still partly denying that sustainability is a real issue or they take pride in undertaking projects. But this is slowly changing.

In order to "spice up" competition in this sector, in 2010 IDH (the sustainable trade initiative), together with several small and medium-sized (SME) spice companies, established the Sustainable Spices Initiative (SSI). SSI aims to develop a globally accepted sustainability standard for spices and help farmers to comply with standards. The global standard is used to compare other sustainability standards and create more transparency on the commitment of companies beyond specific projects.

What is the next phase? Hopefully, the mainstream standard will spark more competitive and commercial interest so that more companies will pay more attention to sustainability. The first aim should be to truly link sustainability to commercial interests and drive sustainability into value chains and into the fields. Then the sector will truly enter the second phase of market transformation.

6.6 Preparing for the next phase

The projects that are started by companies and NGOs in the first phase of market transformation may sound to some like a waste of time and resources. After all, it seems pretty obvious that social projects like training farmers or giving away cooking stoves, or setting up a fire brigade are not the solution to a coffee crisis or devastating forest fires. We all know that the root causes are so much deeper and more systemic. So why bother? The truth is that this first phase is of tremendous value and crucial for further developments. It cannot be skipped. The industry and governments are confronted, for the first time, with the results of their collective action on a large scale. It takes time for them to acknowledge that the problem is real, and it takes time for the industry to get involved and learn. Some companies will start to shift their mind-set from "there is nothing I can do about it" to taking action and actively developing projects. They might even work together with their former opponents, the NGOs. By participating in the projects they will learn what works and what doesn't. This is no small accomplishment.

If done well, this phase will become the foundation for the next phase of market transformation, as the sector gradually learns that there is no easy, quick fix. The next phase awaits: the first mover and competition phase. If done badly, however, the sector will fall back into its old habits, and the momentum for change will be wasted. The sector will remain unchanged, until the next crisis or event presents itself. The characteristics of the first phase of market transformation are summarized below.

Summary of the first phase of market transformation: the awareness and project phase

What triggers change?	- An event, a crisis, a campaign
Main change agents	- NGOs, media, outsiders, visionary individuals - Public pressure is mounting
Who is against change?	- Those that benefit most from the business-as-usual scenario (often the industry, the financial sector, and governments)
Initial response and level of awareness	- Problems are treated with denials, defensive actions, disbelief or people feel that things cannot be changed or it will not make a difference anyway - Problems are misunderstood; only symptoms are visible - If the campaigns hurt enough, initial projects are started
Driving force for the market	- Avoiding reputational damage - Belief in quick fixes - Story-telling and marketing

Willingness to cooperate with others	- There are low-level, confrontational relationships with opponents - There is growing willingness to cooperate on projects with those who have credibility, to share resources, and achieve recognition
How to start this phase and be successful. What can you do?	- Never waste a good crisis: raise awareness of problems - Analyze the situation: find pressure points and target the dominant players in the system; hit them where it hurts - Make your demands actionable - Get the first companies to "comply," claim success, and celebrate your heroes. Put pressure on the laggards
Limitations of solutions and barriers of this phase	- Fragmented and competing projects - Limited and temporary scope and impact - Not scalable; no real exit strategy - The problems keeps coming back as the root causes are not addressed
What comes next?	- Two possible scenarios: - The sector will slowly fall back into its old behavior, until the next crisis Or: - People will slowly become ready for the next phase, where the sustainability issues start to become strategic and competitive

7
The first mover and competition phase

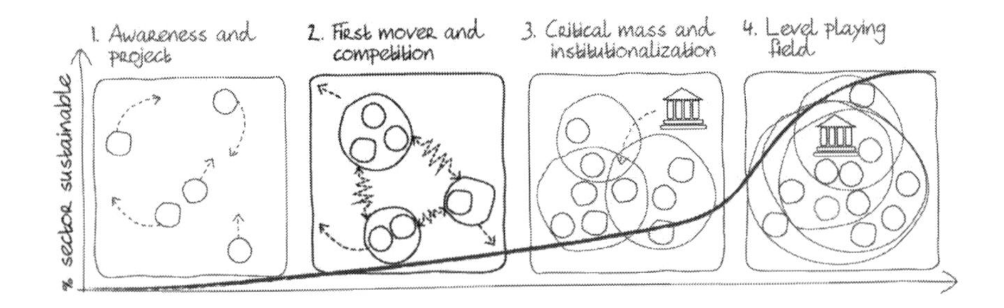

Important conclusions of this chapter:

1. As pressure continues to increase, each sector is learning the valuable lesson that fragmented projects alone do not lead to a structural solution. The sector is now ready for the next phase.
2. First-mover companies start to understand that their responsibility goes beyond their immediate self-interest, but that competitive benefits can also be gained if they play smart and use the situation to their advantage.
3. This triggers a race between companies to score points in the public domain, enhance their reputation and increase brand value. The competition is not limited to companies alone; NGOs, international standards organizations, and others also join the race.

4. Attention for sustainability soon increases, and companies start making public commitments to be a part of the solution.
5. A new industry is born in which sustainability is a hot topic and, for a while, it seems like the solution has been found.
6. However, the problem remains largely unsolved until the industry learns that the root causes have not yet been addressed. More needs to be done structurally to improve the situation.
7. But now, at last, sustainability is on the strategic agenda.

7.1 The sustainability race in the coffee industry

Breaking news! "Douwe Egberts and Utz Kapeh sign an agreement to work together. Douwe Egberts commits to sourcing an increasing amount of certified coffee, starting with 2,500 tons this year!"[197]

On March 28, 2004, newspapers, magazines, and websites celebrated a remarkable step taken by Sara Lee/Douwe Egberts (nowadays called D.E Master Blenders 1753). It was truly a historical moment. After months of careful relationship building, planning, and preparation, Sara Lee/ Douwe Egberts, the third largest coffee roaster in the world and absolute market leader in the Netherlands, agreed to start purchasing part of its coffee from certified sustainable sources. The initial amount of 2,500 tons may seem like a small step for a large coffee roaster like Douwe Egberts, but it was a giant leap for the coffee sector. Stefanie Miltenburg, currently Director International Corporate Social Responsibility at D.E Master Blenders 1753 and Director of the Douwe Egberts Foundation, explained:

> The 2,500 tons of certified coffee was a negligible amount compared to our total volume of 450,000 tons. Nevertheless, it was the very first step taken by Douwe Egberts to stop buying in essentially anonymous markets and move toward traceable and sustainable coffee.[198]

In hindsight, we can see that the public commitment of Douwe Egberts was an inevitable event waiting to happen, sparking the second phase of market transformation: the first movers' competition and the battle between standards.

This historical move by Douwe Egberts in 2004 was a classic result of *urgency* meeting *opportunity*. In the 1990s and first years of the new millennium, Douwe Egberts, just like most of its peer companies, refused to accept direct responsibility for what happened at the farmer level. Most people in the company did not even know what was going on outside their immediate sourcing and marketing practices. Douwe Egberts started to change its views because of the continuing coffee crisis, the ongoing campaigns against the company, and the growing insight that there is no quick fix to this problem. It became painfully clear to the third largest coffee roaster in the world that its sourcing practices matter and that they have an impact on the value chain and on the sector. While Douwe Egberts continued to refuse to work with Max Havelaar/Fairtrade as a matter of principle, the company was nonetheless about to become a leader in sustainability.

Besides the campaigns and the coffee crisis, there was another simple but urgent reason why Douwe Egberts had to act. There was a window of opportunity and the clock was ticking. Other large coffee roasters, who were facing similar campaigns and pressure from NGOs, were starting to act as well. A few months earlier Kraft Foods, an international competitor of Douwe Egberts and proud owner of famous coffee brands such as Maxwell House and Yuban, had signed a partnership agreement with the Rainforest Alliance. The Rainforest Alliance is a network of nature conservation NGOs known for their good work in tropical wood certification. For a few years they had been expanding their work to coffee and other tropical commodity sectors. Annemieke Wijn, Senior Director for Commodity Sustainability at Kraft Foods at the time, looked back on this period:

> At the time, perhaps three people at Kraft knew anything about green coffee production, where it came from or what the production circumstances were, so a partner was absolutely necessary. We looked at the various NGOs that were doing certification, compared their *modi operandi* and how

> we could work with them, and I came up with Rainforest Alliance as the most useful partner for Kraft.[199]

Douwe Egberts knew that Kraft would soon start buying Rainforest Alliance certified coffee, blend it in their products and use it in their marketing. But much more was going on in the industry. Besides the already-mentioned sustainability standards of Fairtrade, Utz Kapeh, and Rainforest Alliance, another initiative was in the making: the Common Code for the Coffee Community (4C). Suddenly, it was getting crowded on the sustainability standards front. The 4C initiative was launched in 2003 by the German Ministry for Economic Cooperation and Development (BMZ), the Deutsche Gesellschaft für Internationale Zusammenarbeit (GIZ; German institute for foreign development cooperation), and the German Coffee Association (DKV). Shortly afterwards, the Swiss State Secretariat for Economic Affairs (SECO), the British Development Corporation, and the European Coffee Federation (ECF) joined in as well.[200] The 4C initiative was not government-led, industry-led, or NGO-led, but was championed by a consortium of various stakeholders in the coffee industry. It invited all kinds of stakeholders (such as coffee brands, traders, growers, NGOs, and other certifiers) to discuss what mainstream sustainability meant for the sector and to agree on a baseline standard for responsible coffee. All the large coffee brands and traders took part in the 4C negotiations—Sara Lee/Douwe Egberts, Kraft Foods, Nestlé, Tchibo, Neumann Kaffee Gruppe, Ecom, Volcafe, but also NGOs such as Oxfam and Pesticides Action Network UK, and the International Labor Union. With their presence and participation, all these companies and organizations endorsed and supported the idea of introducing a common baseline standard for responsible coffee. This was a process not seen before in the coffee industry. It was this process that would spark further competition between coffee companies and enhance the growth of the other certification standards such as UTZ Certified, Rainforest Alliance, and Fairtrade.

The reasoning behind this was simple. When large parts of the industry agree that they will ultimately adopt the same baseline standard, then differentiation in the marketplace is no longer possible. Everybody will, at some point, be doing the same thing. Although this sounds great from a sustainability point of view, it did not make good business sense for the marketing departments. This was one of the reasons why some large roasters did not wait for this to happen, but already began to partner with other standards offering differentiating consumer labels and product claims. This way they could gain first-mover advantages before a baseline standard came into play for all.

7.1.1 Douwe Egberts' grand opportunity

The *opportunity* for Douwe Egberts to become a leader in sustainability presented itself in the form of the Utz Kapeh Foundation. In the Mayan language Q'eqchi, "Utz Kapeh" means "good coffee." In the 1990s the Utz Kapeh Foundation was established in the Netherlands by Nick Bocklandt, a Belgian coffee grower living in Guatemala, and Ward de Groote, former CEO of the Ahold Coffee Company. Utz Kapeh took a different approach to Max Havelaar. It was not founded by NGOs with purely a development mission in mind. Instead it was founded by a coffee grower and a coffee roaster, people from the industry itself who strongly believed that a sustainable coffee farmer is a professional farmer first and foremost. Based on that vision, Utz Kapeh focused more on the implementation of good agricultural practices, in addition to social and environmental sustainability. It tried to stay as close to the conventional coffee trade as possible. Consequently, Utz Kapeh worked with smallholder coffee farmers as well as the larger coffee estates. It certified coffee origins that were mostly used in mainstream blends—not just coffee from Latin America. Moreover, Utz Kapeh did not believe in working with minimum prices. Instead, it worked with the principle that *a better product deserves a better price*. This was more in line with regular market principles. Moreover, Utz Kapeh strongly believed that the emphasis in marketing should not be on the certification hallmark prominently displayed on the packaging. Instead, sustainability should be an integral part of the brand value and positioning. Utz Kapeh simply provided a credible endorsement of that brand message.

What are sustainability standards and certification programs?

In many of the agricultural commodity markets, sustainability standards and certification programs have proven to be a very effective way to link companies with sustainability issues in their supply chain. But what are they and how do they work?

What are sustainability standards?

Standards are lists with field practices and indicators that define sustainable (socially and environmentally responsible) production. They tell the farmer what to implement and how to produce in a more sustainable way. Examples of these practices are: children have to go to school; pickers and seasonal workers need to have a contract and need to be paid a minimum wage; effluent water needs to be purified and treated before it is discharged; waste needs to be disposed of properly; tropical forests cannot be cut down. The idea is that implementing these practices not only makes farmers more sustainable, but also better organized and more productive.

Every year an independent verifier checks if the farmer actually complies with these practices and, if he does, the farmer is certified. This means he can sell his product not as a normal undifferentiated product, but as a certified product. This can lead to better trade terms for the farmer: a better price, a long-term contract, better relationships with traders, and access to more markets.

The buyer who buys the certified product can make a claim on its product in the form of a label showing the consumers or customers that they are sourcing a sustainable or responsible product.

For whom are standards valuable?

Due to pressure from campaigns or from a competition point of view, particularly in the second phase of market transformation, many companies and brands are looking for sustainability assurance and will want to buy a certified product. There are two levels of value creation for them:

1. Independent assurance that the product they source is produced sustainably and that certain bad practices no longer occur. This protects their reputation and ensures they are not attacked for non-sustainable practices. This is the insurance component of the standards.

2. They can use certification in their marketing and can make claims by using certification hallmarks on their products or on their websites and brochures. This leads to added brand value.

For many NGOs, certifications are attractive because they want to implement standards at farms to improve the environment and social conditions. Moreover, the farmers will receive better terms of trade which can be in the form of a better price or a better relationship with their buyers. The NGO can claim that they helped the farmer to achieve that.

For governments and campaigning NGOs, standards are an effective tool because they are an easy way to measure and verify if the industry is living up to its commitment.

In short:

Standards create a link between responsible buyers and sellers, and they offer value, both for farmers and brands, to implement better sustainability practices in the field. Standards therefore seem to be the ideal tool to change market behavior and to tackle sustainability problems.

These differences are nothing special nowadays, but back then this was a new way of thinking. Indeed, it was quite revolutionary, and these more market-driven principles were much more compatible with industry giants such as Douwe Egberts, Paulig, Migros, and other coffee roasters, retailers, and traders. The idea of mainstreaming sustainability started to catch on.

However, in the early days of the foundation the credibility of Utz Kapeh itself became an issue due to its origin in the coffee industry. After years of effective campaigning against this industry by NGOs, an initiative from that same industry was not immediately believable. In fact, many NGOs considered Utz Kapeh to be a watered-down version of Max Havelaar/Fairtrade and a cheap strategy for the industry to avoid joining Fairtrade which was in their eyes the only true solution. As a result, some NGOs considered Utz Kapeh to be a threat, and it did not take long before the foundation itself became a target of NGOs aiming to discredit or question the initiative.

The resistance against Utz Kapeh from NGOs was potentially a serious threat. This changed after the founders made an important decision. They decided to transform the organization into a foundation that was independent from Ahold (the founding retailer), and invited Nico Roozen from Solidaridad to join the board of directors. As mentioned in Chapter 6, Roozen was no stranger to the coffee industry. He had, together with Frans van der Hoff, founded Max Havelaar almost 25 years earlier in

1988.[201] As such, Roozen was not only a pioneer in market-based development solutions in the coffee sector, he also knew the NGO sector well. According to Roozen:

> In 1998, ten years after the founding of Max Havelaar, I concluded that it had very limited impact and potential, while my ambition has always been to go for mainstream scale. I knew at that time that Max Havelaar would not change and innovate.[202]

Roozen believed that the Utz Kapeh concept, with its more mainstream features, might be able to change the industry.

Roozen's endorsement of the Utz Kapeh concept initially led to even more resistance from NGOs against Utz Kapeh, as many NGOs saw this as a betrayal by a founder of what, in their eyes, was the "true cause." On the upside, it gave the coffee industry an important argument to view Utz Kapeh as a credible, new-generation concept that could safely be embraced. Becoming an independent foundation also meant that the first donors could get on board. This was absolutely critical for Utz Kapeh's success.[203] The foundation was now open for business and ready to become a credible partner for one of the industry leaders, Douwe Egberts.

7.2 On your marks ... get set ... go

It was like a racing engine starting. The public commitment by Douwe Egberts in 2004 started an unprecedented international competition on sustainability. It didn't take long for Kraft Foods to respond. Sustainability in coffee, which for years was a topic to ignore, avoid, or deal with through projects and charity, was suddenly the focus of fierce brand competition. National and international coffee brands were starting to compete on who was the most sustainable and were fighting over who could gain most recognition from consumers, clients, and NGOs.

A few months after Douwe Egberts publicly committed to source UTZ Certified coffee and blend it into the mainstream brands, the company launched a special 100% certified "out-of-home" coffee product that was sold to businesses exclusively. Kraft Foods quickly followed with a similar public commitment to buy an increasing amount of Rainforest Alliance certified coffee. Kraft's initial committed volume was, of course, slightly bigger than Douwe Egberts. This in turn led to other brands stepping up their efforts and increasing their public commitments as well, which forced others to react, and so forth.

NGOs, in the meantime, quickly claimed success, celebrated their heroes, and put extra pressure on laggards. Coffee roasters and retailers everywhere were now forced to take sustainability seriously. The boards of the companies had to make some strategic decisions. What should we do? Do we follow Douwe Egberts with Utz Kapeh? Follow Kraft Foods with Rainforest Alliance? Should we go with Fairtrade perhaps? Should we put our money on the 4C baseline standard instead and claim that we believe in an industry-wide approach? Or can we afford to do nothing and still avoid negative press and campaigns? Some of the more exclusive coffee brands, such as Nespresso from Nestlé and Starbucks, took a different position altogether. They decided to create their own standard and verification system instead. This, they believed, was much more compatible with their exclusive and unique marketing position.

The game was on. A better sustainability profile for the coffee roasters was the goal, and the market was the battleground. In only a few years, Douwe Egberts' initial volume of 2,500 tons of certified coffee per year grew to more than 25,000 tons per year. Other roasters made similar commitments, even publicly claiming they would ultimately reach 100% certified volume. Those were the days.

The competitive game was not limited to commercial companies. NGOs also took part. Just as companies embraced different labels to differentiate themselves in the marketplace, so did the competing NGOs. These NGOs easily established partnerships with big companies by partnering with one of the certifiers. As mentioned before, Solidaridad embraced Utz Kapeh from the very beginning, and it did not take long before they set up "the coffee support network," a capacity-building program to help coffee farmers become compliant with the Utz Kapeh standard. Solidaridad partnered with Douwe Egberts and other coffee brands to start projects and programs. Oxfam became a founding member of the 4C, despite its open preference for the Fairtrade system. This was an important step for the 4C, and it used the Oxfam presence to support its legitimacy. Other NGOs swore allegiance to the Rainforest Alliance standard, partnering with brands that used the Rainforest Alliance certification system such as Kraft Foods and other coffee brands, mostly in the U.S. For NGOs, this was no small game. A lot of money was at stake. Helping farmers comply with sustainability standards is not cheap or something that can be done quickly; it therefore requires vast, multi-year, multi-region, multimillion-dollar programs. Money makes the world go round, and this is the case particularly when you are an NGO and a new, attractive, industry is emerging.

The sustainability certifiers themselves quickly followed the competition mind-set. Each of the certifiers was completely convinced that their particular standard was fighting for the right cause and they were making a real difference to the lives of many coffee farmers. The case for change was obvious. Coffee farmers, who implemented their standard prescribing better and more sustainable farming practices, become better, more sustainable farmers. With large buyers looking for certified coffee, these farmers had better relations with the large international markets, and they were paid a higher price for their coffee. This was another benefit for the farmers. The large multinationals on the other side of the value chain received a better product and they were "willing" to pay a higher price for the certified coffee because, in turn, they could place a label/claim on their brands and coffee packages. This was the added value for the brands. It was a perfect value cycle for everyone.

All of this was "credible" and "real" because external auditors verified the compliance of the farmers with the standard and because NGOs endorsed the standards, so they must be doing something right. It was a win–win situation for everybody. Each of the standards and certification programs was encouraged by their own groups of clients, NGOs, donors, boards, and farmers to grow in volume, get more clients, and to compete with other standards. After all, when you are *a* standard it is only natural that you want to become *the* standard.

The big companies such as Douwe Egberts, Nestlé, and Kraft chose their standard, but the smaller players did so as well. For example, Tchibo, a German retail chain and seller of specialty coffees, partnered with

Rainforest Alliance and Fairtrade (and, years later, with Utz Kapeh as well), while the Swiss supermarket chain Migros chose Utz Kapeh. In response, the second largest Swiss retailer, COOP, partnered with the 4C. Competition started even in Japan, where the main driver was not so much NGO campaigns or concerns about sustainability, but concerns about food safety. In 2006 Utz Kapeh introduced its traceable, food safe coffee in the Japanese market, and the Rainforest Alliance followed soon after. Vermont-based Green Mountain coffee focused on Fairtrade, while Paulig, a Finnish coffee giant, chose Utz Kapeh. And so the story continued.

Meanwhile, most of these coffee companies were also taking part in the further development of the 4C baseline standard. Carsten Schmitz-Hoffmann, one of the senior managers of the 4C initiative at the time, looked back on this period:

> Personally, I always understood the 4C as a stepping stone approach for producers to get into the other more demanding certification systems. But the other standards did not necessarily accept us in this position and saw us as competition. That was a phase of very intensive discussions.[204]

7.2.1 Coffee standards gain in popularity

The standards created a new industry. Sustainability was hot, and the industry was competing on this issue. NGOs benefited from this competition, as more money than ever was being pumped into the sector. The standard and certification business peaked in popularity in 2010 when the Dutch coffee industry led by the Royal Dutch association of coffee roasters and tea packers (the KNVKT) signed a declaration that the Dutch coffee roasters as a whole committed to sourcing 75% of all coffee consumed in the Netherlands as certified sustainable by 2015[205] (in 2013 the market share of sustainable coffee was 50%).

For the Dutch government, as well as governments elsewhere, the standards industry was a blessing. They now had an easy way to push the coffee industry to be more responsible and, at the same time, they had an easy way to verify if the industry actually complied. They only had to check the volume of certified coffee that was being sourced and, as can be seen in Figure 11, this volume was growing exponentially.

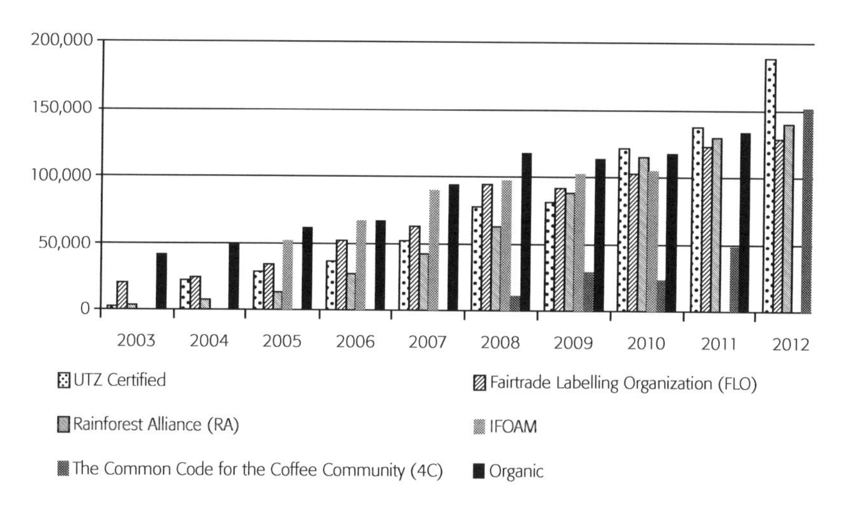

FIGURE 11 Sales of certified coffee per certification standard in metric tons (MT) between 2003 and 2012[206]

After the commitment of the Dutch government, the general feeling was that coffee was "done." The coffee sector had found the golden solution to its problems by implementing sustainability standards. From now on it just needed to increase the amount of certified coffee sold, and then everything would be OK.

This would later prove to be a false sense of security. Coffee was not "done"; despite all the successes, public commitments, and declarations, the coffee sector was only at the end of the beginning of its true change process. It would turn out that implementing a sustainability standard for the first 10% of the coffee farmers (the better-organized ones) was easy compared to the next 10% to 20% of the less organized farmers. Certifying these farmers proved to be a costly exercise that the companies and NGOs were paying for. Moreover, the added value that once drove competition between the companies slowly diminished. Since almost every self-respecting coffee company now had committed itself to go for sustainability, the differentiation factor diminished, and the added value decreased, even though costs continued to rise. Consumers took it for granted and the media paid less attention to the big public commitments of companies. It became normal. According to Nico Roozen, Managing Director of Solidaridad: "When a large percentage of the coffee or chocolate products on the supermarket shelves in the Netherlands have a logo or hallmark, the importance of the label diminishes."[207] And when there is no added value, but only added costs, the companies lose interest, while the problems persist. This shows the limitations of a model based on competition alone.

On top of that, the first doubts were starting to emerge and critical voices were saying that the implementation of standards alone was not the silver bullet (i.e., that it did not lead to real, lasting impact on the ground and that it would not be possible to certify all the farmers). The fierce competition between the coffee companies had driven a rapid uptake of sustainability standards in the market. But, as we saw in Part I, a system failure cannot be fixed with standards alone. More needed to be done to fix the real root problems of the sector.

In the words of David Rosenberg, Group Sustainability Advisor at Ecom Agroindustrial Corp and former CEO of UTZ Certified:

> Certification does not equal sustainability. Certification is a first step, not a last step. The problem with certification and compliance is that it can stifle innovation: once you have qualified, there is little reason to do more. It is an important first step, a common language that you can build on. It remains a question whether it is the tool to continue the market transformation.[208]

After years of competition on standards, labels, and public commitments, the sector slowly prepared itself for the next phase of market transformation: critical mass and institutionalization. In the coffee sector, however, it would take a few years before the industry was really ready to take the next step.

7.3 The sustainability race in the cocoa sector

As we have seen in the previous chapter, the cocoa sector was no stranger to tough sustainability challenges nor to NGO campaigns. But, in 2005, the real fireworks were still to come. By then the coffee sector was already showing so much progress and impact , and it was just a matter of time before this dynamic spread to the cocoa sector. This indeed did happen, albeit with some delay.

After years of campaigns to raise awareness on the issues of child and forced labor, the game changed when the Harkin–Engel Protocol was signed in 2001. The protocol was the result of years of intense campaigning and media reports about alleged malpractices in cocoa production. The objective of Harkin–Engel was to force the cocoa industry to abandon the worst forms of child labor and slavery. However, it didn't go quite as planned. Results were lacking and the industry was losing credibility in the eyes of the public and the U.S. government. Companies were now on the lookout for a new approach.

7.3.1 Let the games begin

Not long after the deadline of the Harkin–Engel Protocol, Nestlé partnered with the Rainforest Alliance in 2005. In that same year, two of the largest and leading cocoa processors and traders started their own programs as well. Barry Callebaut started the Quality Partnership Program and ADM started the Socially and Environmentally Responsible Agricultural Practices (SERAP) program. Cadbury, a large UK-based confectionary company, took over Green & Black's in 2005, a premium organic chocolate brand. Later, in 2008, Cadbury would start the much bigger Cadbury Cocoa Partnership program. And, one year later, Cadbury announced that its famous Dairy Milk chocolate bar would become 100% Fairtrade. As *The Guardian* reported at the time:

> Dairy Milk is the first mainstream chocolate bar in the UK to be sold with a commitment to pay cocoa suppliers the "Fairtrade premium" of $150 (£105) a tonne above market prices. When the bars go on sale this summer the value of Fairtrade chocolate sales in Britain will leap from £45m to £225m [$72 million to $360 million]. Cadbury's pledge to buy 10,000 tons of cocoa under Fairtrade terms will triple certified sales from Ghana.[209]

This was an amazing move by a mainstream cocoa company, and it did not stop there. Royal Verkade, a Dutch subsidiary of United Biscuits, and the absolute market leader in the Netherlands, announced that all its chocolate would be 100% Fairtrade-certified by September 2008. This was another big success for the Fairtrade system.

UTZ Certified (Utz Kapeh had changed its name to UTZ Certified in 2007), launched its cocoa certification program in 2007. It was an immediate success and almost instantly became the largest industry-endorsed cocoa certification program. This showed the urgent need of cocoa companies for a credible, practical, and mainstream third-party certification program. Within a matter of months, several large cocoa brands, traders, and leading NGOs such as Nestlé, Mars, Cargill, Ecom, WWF, and Solidaridad joined the UTZ Certified cocoa program. This consortium of companies and NGOs started to develop and implement a mainstream sustainability standard for cocoa production. Solidaridad, the Dutch NGO that played such a crucial role in the success of UTZ Certified in coffee, immediately saw the opportunity and started large field programs in cooperation with industry partners to help cocoa farmers become certified. By the end of 2009, about 3,600 cocoa producers had been trained to produce sustainably and became certified. In November 2009, the first 5,000 tons of UTZ Certified cocoa reached the port of Amsterdam. By mid-2011, around 100,000 UTZ Certified farmers were producing over 70,000 tons of certified cocoa a year.[210] The UTZ cocoa program remains one of the largest certification cocoa initiatives to this day.

A further acceleration of the certification and field implementation program came in 2008 when IDH, the sustainable trade initiative, joined the program. IDH was founded in 2008 by several Dutch Ministries with a mandate and funding to help accelerate and scale up sustainable trade initiatives. One of the first actions of IDH was to "adopt" the UTZ Certified and Solidaridad cocoa program and boost its scale with large financial injections.

Matthieu Guémas, former Corporate Responsibility Manager at Cargill and currently Head of Business Development for Europe, Africa, and Asia at GeoTraceability, looked back at this period: "In 2008 we saw a shift in mind-set in the cocoa sector. We moved from sustainability as a very limited niche market with some activities, to: 'Ok, what if we apply that to the larger mainstream?'"[211]

In April 2009, Mars committed to "fundamentally changing the way sustainable cocoa farming practices are advanced by aiming to certify its entire cocoa supply as being produced in a sustainable manner, by 2020."[212] Alastair Child looks back at this period:

> Whilst external market pressures had brought standards and certification into the spotlight, Mars' primary concern was farmer livelihoods and supply security at origin; hence our first move was a 100% supply chain commitment. We realized the power of standards was the potential for scale, as they leverage the only network that reaches every cocoa farmer in the world—the supply chain.

Mars's very first concrete action was its announcement to source 100,000 tons of Rainforest Alliance certified cocoa per year, by 2020, starting with its iconic Galaxy Bar in the UK.[213] "It is the appropriate choice for a stable, high-quality cocoa supply in the future," said Howard-Yana Shapiro, Global Director of Plant Science for Mars. "We think that this is what we are supposed to do as a responsible world citizen."[214] The move by Mars was the trigger for other companies to react. That same year, Nestlé, who was already a partner in the UTZ Certified program, took its commitment to a sustainable cocoa sector a step further. It announced its CHF110 million ($110 million at that time) ten-year Cocoa Plan to support cocoa farmers and their communities.

The WCF, originally founded in 2000 by the cocoa industry, also began to scale up its sustainability programs. In 2009 it launched the Cocoa Livelihoods Program (CLP), a five-year, $40 million initiative funded by the Bill and Melinda Gates Foundation and 15 chocolate companies. The goal of the program was to significantly improve the livelihoods of approximately 200,000 cocoa farmers and their families in Cameroon, Côte d'Ivoire, Ghana, Liberia, and Nigeria. In 2011, WCF launched an additional program called the African Cocoa Initiative (WCF/ACI), which was co-funded with USAID. This was a $13.5 million five-year public–private partnership program aimed at institutionalizing effective public- and private-sector models to support sustainable productivity growth. It also aimed at improved food security on diversified cocoa farms in West and Central Africa.

7.3.2 National industry commitments

The proliferation and competition of sustainable company programs and certification commitments was now starting to reach its peak. Inspired by the public commitment of the Dutch coffee industry, the Dutch government now rallied the Dutch cocoa industry to do something similar.

This was a significant step. Beginning with its colonial past, the Netherlands has been the largest importer and processor of cocoa beans and the largest exporter and re-exporter of processed cocoa and chocolate in the world. In 2010 the Dutch cocoa industry, the Dutch Ministry of Economic Affairs, and other stakeholders signed a declaration of intent whereby the Dutch cocoa sector (industry, government, and civic organizations) committed to 100% certified cocoa consumption by 2025. This was an absolute triumph for the certification programs and their standards. Now two major commodity sectors, the coffee and cocoa sectors in the Netherlands, had publicly committed to source large amounts of certified sustainable products.

7.3.3 Perception versus reality on the ground

However, the reality in Côte d'Ivoire and Ghana was totally different from the reality in the boardrooms of the companies that had signed the declaration and were setting up the large improvement programs, and buying increasing amounts of certified cocoa. Despite the innumerable programs that had started and the hundreds of millions of dollars that were being spent, the problems of child and forced labor were not being eradicated and farmers' productivity and income was not being sufficiently impacted. The situation was complicated by the outbreak of the second civil war in Côte d'Ivoire in 2011. This made it close to impossible to implement sustainability programs while people were shooting at each other and staff were being evacuated. This showed again that the reality on the ground is different from marketeers' claims and product labels.

Despite the ongoing civil war, in 2011 CNN aired another documentary on child slavery and announced a boycott of Hershey.[215] Hershey is the largest chocolate manufacturer in North America and this campaign put the issue of child and forced labor back into the public eye. After rejecting certification for years, Hershey now had to respond, and it decided to join the sustainability and certification circus. It announced its 100% Rainforest Alliance certified Bliss chocolate bar and, in early 2012, Hershey pledged $10 million to educate West African cocoa farmers on improving their trade and combating child labor.[216,217]

The story continued. In 2012 Mondelēz International (formerly part of Kraft Foods), owner of famous chocolate brands such as Cote D'or, Milka, and Toblerone, launched an unprecedented ten-year sustainability program with a $400 million budget to support 50,000 farmers and their cocoa growing communities. That same year, Cargill, the largest cocoa processor and exporter in the world, launched its massive Cocoa Promise program, in which it pledged to help 60,000 farmers to become certified and to provide educational opportunities for 22,000 children.[218]

Sustainability in the cocoa sector was booming. Never before was there so much attention for, and competition on, sustainability. Certification programs were at their peak. But more critical voices were being heard. Were they really making an impact? Was all of this really going to change the industry?

7.3.4 Reality strikes

Meanwhile, in the Netherlands, the Tropical Commodity Coalition (TCC) was formed by Dutch NGOs who had campaigned so successfully against Douwe Egberts at the time. The self-proclaimed mandate of TCC was to act as the tea, cocoa, and coffee industry watchdog, making sure the industry lived up to all its claims and commitments. To this end it published an annual report, the *Tea, Coffee and Cocoa Barometer*, on the state of certification in relation to the public claims of the major companies in the coffee and cocoa sectors. This was an effective strategy. The report quickly gained influence as it visually displayed the yearly progress of the companies in sourcing certified coffee and cocoa. In 2010, it started to criticize the use of certification as a tool to make cocoa farming more sustainable:

> Without proof of impact, standards and sustainability programs are in danger of being questioned. More and more farmers, consumers, companies and governments show interest in the results of sustainability initiatives. The ways in which we now measure results lack transparency. Nor is there agreement on what sort of results need to be measured.[219]

Even leaders of the cocoa industry started to openly question the impact of certification. According to Nicko Debenham, former Head of Cocoa at Armajaro and now Head of Sustainability at Barry Callebaut:

> I think that what needs to happen is for the certification schemes to start measuring their success or failure on the impact at farm level and on farmer livelihoods, rather than on numbers of farmers trained, volumes certified, and market shares. We have seen a lot of certified cocoa entering the market, and buyers paying a premium price for it. Unfortunately, this so-called certified cocoa is not generating any impact on the ground, either because farmers have not received the premium in cash or in kind, and/or because good agricultural practices have not been implemented.[220]

Daudi Lelijveld, formerly working for Cargill and also former Vice President of Barry Callebaut's cocoa sustainability programs, stated:

> In my humble opinion, offering different and often conflicting programs has become unhealthily competitive. I think the world, and more importantly the farmers, are worse off through this competition between standards. I'm referring to the proliferation of all kinds of certification and standards and the standard-raising bodies—the process and awarding of certificates has become a business in itself. I am a great believer that an industry should not have 15 standards for one industry but should have one standard which is managed, controlled, and evolves in the hands of the industry, with all its stakeholders represented.[221]

So here we are, years after the deadline of the Harkin–Engel Protocol has elapsed. Dozens of company programs, multi-stakeholder initiatives, sustainability standards, and overarching initiatives have emerged. Hundreds of millions of dollars have been spent and hundreds of thousands of farmers have been trained to become certified. Premiums were paid for certified cocoa, and products have been labeled with promises and claims. Meanwhile, however, the real problems of child and forced labor, poverty, and low production have continued at unacceptable levels in the sector and are still hurting the industry. Moreover, cocoa was growing scarce; trees and farmers were getting older; and a future decline of cocoa production was unavoidable. Slowly, minds in the cocoa industry began opening to the idea that competition on sustainability is, perhaps, not the whole answer. It leads to fragmentation, which in turn leads to inefficiencies and ineffectiveness. If something is going to be done about this, a change in strategy is needed to really tackle the roots of the problem:

instead of competition on product labels, collaboration with competitors and governments is needed. This insight marks the end of the second phase of market transformation. The next phase is that of critical mass and institutionalization.

7.4 The first mover and competition phase in palm oil

The massive fires of 1997 incinerated a huge area (the size of Switzerland) of Indonesia's tropical forests, and polluted the air over much of South-East Asia. What followed were several large and effective campaigns to raise public awareness and to put pressure on mainly Dutch banks as one of the main drivers of the unsustainable expansion of palm oil estates.

In the years immediately following the fires, the WWF successfully established several partnerships with large industry players, such as Unilever and Migros, on the issue of palm oil. Building on its strong track record and experience with setting up other important multi-stakeholder initiatives and certification standards—such as the Forest Stewardship Council (FSC), the Marine Stewardship Council (MSC), and many others—in 2002 the WWF took the next step and held a multi-stakeholder meeting in London. Representatives of retailers, food manufacturers, palm oil processors, traders, and WWF came together for the first time to discuss the outlines of a sustainable palm oil sector. This was the beginning of what later would become the Roundtable on Sustainable Palm Oil (RSPO),[222] the sector certification program for sustainable palm oil. There was a lot of pressure on this roundtable meeting, as the forest fires that had raged in 1997 returned that year, and once again were making newspaper headlines. This time the fires were in Central Kalimantan, Indonesia, and the evidence again showed that more than 75% of the fires originated from palm oil plantations, timber plantations, and forest concessions.[223]

With the convening power of the WWF, and an increasing shared sense of urgency to really make a change, a "coalition of the willing" began to form. After the initial meeting in London, the first roundtable was held in Malaysia in 2003 and also included producers. The initiative was formalized one year later as the RSPO.[224]

According to Darrel Webber, Secretary General of the RSPO:

> People only change when they see themselves on a burning platform. In the case of palm oil this was almost literally the case. It was the regional haze, which blanketed most of South-East Asia and certain parts of Asia in general, that started to attract a lot of attention. Combine that with the results of research at that time, the run-up to the defining of the Millennium Development Goals, the improvement of internet connections and satellite imagery to show biodiversity loss and loss of forests. I think all of these things combined helped to create a perfect environment for the establishment of the Roundtable on Sustainable Palm Oil.[225]

After formally setting up the internal governance of the RSPO, the first big task for the participants was to agree on the definition of sustainable palm oil in the form of a standard.[226] For this task a technical working group was formed with representatives from all stakeholder groups. Understandably, the divergent interests made this no easy process. Webber looked back:

> In the beginning it was very, very tough. The seven stakeholder groups from the supply chain, which includes environmental and social NGOs, came together and basically no one trusted each other. There was lots of heated debate, lots of arguments. It took more than a year to draft the first standard. Many close calls, with threats of walkouts and dissatisfaction. But at the end of the day, trust was built over time. People could start to understand the other parties' views much better over time.[227]

It was also becoming clear that the slow process would not be producing any fast results that would impact the reality on sites where palm oil was planted or grown. Jan Maarten Dros described these meetings: "Only talking and drafting a standard would not change the reality on the ground, even if it was in the right multi-stakeholder setting."[228]

In 2005, three years after the RSPO was founded, the members finally adopted the principles and criteria that had been proposed. For proper testing and empirical validation, a two-year pilot phase was scheduled with volunteer firms. The idea was that these pilots and validation would eventually result in the very first sustainability standard for palm oil.[229] In the end, it would take three years to get the first certified sustainable palm oil sold on the market. The RSPO still had teething problems and received criticism. Several companies had experimented with RSPO, but found it to be complicated, costly, and hard to implement.[230] It was also found to be non-inclusive, as smallholders could not afford to adhere to the standard and adopt best practice, and it was criticized for not addressing the real issues effectively.[231]

Darrel Webber looked back at this period:

> We found it very difficult to get the message out to the world of farms, to make the business case for the producers and to convince them to change. Because, at the time, the palm oil industry was faceless. Nobody knew them and there was no brand attached to palm oil. There was nobody pressuring them to do anything. That was difficult in the beginning, to get more people on board.[232]

Despite the good efforts of the coalition of first movers, critical voices were never far away and skeptical NGOs kept campaigning, not giving the food companies any rest. In 2005 Friends of the Earth published a report, *The Oil for Ape Scandal: How Palm Oil Threatens Orangutan Survival,* about the impact of palm oil plantations on the orangutan,[233] and a campaign followed in the UK.[234] A year later, another study showed that implementation of the anti-deforestation policy of Dutch banks was still not complete and that the policies were at risk of being "rendered useless."[235] The banks, still remembering the effects of the campaign against them a few years earlier, reacted quickly by announcing the launch of the Equator Principles, a common approach to managing environmental and social risk associated with financing large projects.[236] To ensure that the banks would finally live up to their words a new NGO, BankTrack, was formed. BankTrack functioned as a watchdog on the banks' compliance with the Equator Principles.[237] The message to the banks was clear: we are not going away and there is no escape.

Frustrated with the lack of progress, NGOs continued putting pressure on the industry. Greenpeace stated that:

> The members of the Roundtable on Sustainable Palm Oil (RSPO) represent 40% of palm oil production and trade. Currently, it is not sufficiently robust to address deforestation and peat land degradation in Southeast Asia. ... In numerous cases (certified), RSPO producer members are still establishing plantations in peat lands or High Conservation Value forest areas.[238]

As a result, in 2009 Greenpeace activists dressed as orangutans stormed Unilever's headquarters in London with banners that read: "Unilever: Don't Destroy the Forests." The campaign hurt Unilever, as it was also chair of the RSPO. Unilever later announced it would source all its palm oil from "sustainable" sources by 2015.[239] Even more admirable and iconic, it also cancelled a €30 million contract with Golden Agri-Resources (GAR), the palm oil branch of Sinar Mas. The latter is a large Indonesian chemical and lumber products conglomerate which had been accused by Greenpeace for years of illegal and unsustainable palm oil expansion, illegal deforestation, and involvement in the destruction of peat land, rainforests, and orangutan habitat.[240] A year later, Nestlé would also break its contracts with Sinar Mas after facing the same pressure from Greenpeace.[241] Finally, sustainability was becoming serious business in the palm oil sector.

7.4.1 Governments (re-)enter the arena

With all the dynamics in the sector, national governments slowly assumed more responsibility as well. In 2007, the Malaysian Palm Oil Council (MPOC) launched a small $5.5 million revolving fund to improve sustainable practices and biodiversity conservation related to palm oil production.[242] And in 2008, the Indonesian government started the development of its own sustainable palm oil standard: Indonesian Sustainable Palm Oil (ISPO), which was launched in 2009. The ISPO standard was designed to ensure that all Indonesian oil palm growers, including smallholders and medium-scaled plantations, adhere to sustainable palm oil management and comply with Indonesian legislation.[243] In 2011, the Indonesian government took a big step and imposed a moratorium on the conversion of primary forests and peat forests.[244]

Many people in the palm oil industry asked why another national standard was necessary, when the RSPO system was already in operation. As Jan Maarten Dros argued:

> When two countries produce more than 75% of palm oil traded on the global market, every interference with the production, financing and trade of that commodity is seen as an infringement on their development strategy and sovereignty. Indonesia developed ISPO as a response and the government supports this by saying: "Our mainstream industry can now distance itself from the RSPO that serves all kinds of international and foreign interests that are not the Indonesian interest. Our ISPO standard is highly credible because we as a government guarantee it, so any company can buy ISPO oil instead of RSPO oil."[245]

7.4.2 Increasing competition and gaining momentum

With the emergence of the ISPO, the palm oil sector also entered the second phase of market transformation, that of first mover and competition. In this case, the competition was driven less by brand value, product labels, and consumer preference, and more by political and national sovereignty issues. Jan Kees Vis, Global Director of Sustainable Sourcing Development at Unilever and former chair of the RSPO, looked back on this period:

> The RSPO was a bit nervous when the ISPO was announced. But at the end of the day you have to put it in perspective: in 2002 nobody in Indonesia spoke about sustainability. Now, the Indonesian government has created regulations for sustainable palm oil themselves. We must see that as a great step forward. The discussion has matured: it is no longer on whether we need sustainability or not but on how much sustainability we need.[246]

Then another sustainability standard entered the arena, the International Sustainability and Carbon Certification (ISCC). The ISCC was developed in 2010 with the support of the German Federal Ministry of Food, Agriculture and Consumer Protection. The aim of this standard was to regulate and promote sustainability for biomass used to produce biofuels and bioenergy. The ISCC focuses on the reduction of greenhouse gas emissions, the sustainable use of land, the protection of natural bio-spheres, and social sustainability.[247]

Gradually, Western national governments also began to expand their role and set clear industry targets. The Dutch government made the first move, excluding palm oil from its national green energy subsidy scheme in 2008.[248] In 2010 it pledged to import 100% sustainable palm oil by 2015.[249,250]

7.4.3 Has it all made a difference?

As a result of the continuous campaigns against multinationals based in Europe and, increasingly, in the U.S., with the appearance of sector-wide and national sustainability standards and the ever-increasing participation of governments, the demand for sustainable palm oil began to grow. The RSPO and the movement toward a more sustainable palm oil sector gained significant momentum. More and more companies joined the RSPO and made commitments, just as we have seen in other sectors. Companies such as Blommers, a large chocolate manufacturer, Starbucks, and McDonald's announced their commitment to sourcing sustainable palm oil.[251,252] Ferrero also announced that Nutella would switch immediately to 100% sustainable and traceable palm oil.[253]

Bruce Wise, Sustainable Business Advisor at International Finance Corporation (part of the World Bank group), illustrated the usefulness of standards:

> The sustainability standards have been important for the relationships that have been established by a multi-stakeholder process. If you had said five years ago that WWF would be sitting with Cargill and JBS on a beef roundtable, people would have laughed at you. If you had said Greenpeace would strike an agreement with GoldenAgri and Tropical Forest Trust and would present the work that they're doing to support these companies at international conferences, I don't think people would have believed it. But here we are.[254]

All these market dynamics have ensured that, in only five years after its roll-out, the RSPO standard currently certifies around 14% of global production of palm oil.[255] This is a huge success.

However, despite the large growth in membership, industry commitments, and market share, the issues in the sector remain: land grabbing, degradation of peat lands, and the poor position of smallholders. Also, the fires came back, this time worse than ever. In June 2013, after another period of drought, the forests caught fire again and the choking ash clouds returned. Satellite images reported more than 700 severe fire alerts. Then, only a few months later in September 2013, another fire season followed and then again, in March 2014, large forest and peat areas were set on fire.[256] This time satellites revealed more than 1,400 significant fire sites, and nearly 50,000 Indonesians were reported to have suffered severe respiratory ailments due to air pollution. The fires were extensive in areas with deep peat soils, suggesting that high volumes of carbon were being released, contributing to climate change. Sadly, many fire sites were yet again linked to palm oil estates and forestry companies.

Since the fires continued to come back, questions were being raised about whether certification was addressing the right problem. Greenpeace accused RSPO members of being actively involved in causing the recent forest fires, which had resulted in Asia's worst air pollution crisis in decades. As evidence, Greenpeace released maps showing fire hotspots on land owned by member companies of the RSPO. According to Greenpeace: "The RSPO has failed to tackle its members' role in creating the conditions that led to such a disaster, nor has it held companies accountable for the impact of their operations."[257]

The recent fires were so severe that the Indonesian government was forced to take steps. In June 2013, after many years of delay, Indonesia agreed to become the final member of a South-East Asian anti-pollution pact. The media saw this as a hopeful sign that the Indonesian government was finally accepting its responsibilities.[258] Unfortunately, it won't publicly reveal precisely where timber and agriculture companies operate in the areas where the fires occurred. This gives little hope for change.

7.4.4 The palm oil sector is getting ready for the third phase

It has been more than 15 years since the massive forest fires in 1997 alerted the world to the palm oil problem for the first time. Since then the sector has been in turmoil. Countless campaigns have impacted the industry. Many well-known companies have taken responsibility and publicly announced their commitments to source only sustainable palm oil. But the constantly returning massive forest fires are a dark and gloomy reminder that the issues have not been solved, and that more than certification and different sourcing is needed to change the sector.

Why standards do not work: a plea against the use of standards

Besides the arguments in favor of the use of standards mentioned at the beginning of this chapter, there is a growing case against the use of standards as the main instrument of change, as their impact and reach is limited and their costs too high.

Why do standards not always work? They are too difficult and too expensive

Standards are lists of field practices that define sustainable production. It is not uncommon that these standards are documents with hundreds of criteria with which farmers need to comply. Implementing and complying with all these criteria is very costly for the farmer, requiring, in many cases, the implementation of all kinds of new practices and techniques. Often the farmers do not know, and do not understand, these new techniques; they therefore need extensive training.

In order for the verifiers to check compliance with the new practices, farmers need to implement internal control systems (ICS). These systems require the farmers to keep records of what they did in the fields, and how and when they used certain inputs. Implementing and using all these new techniques and internal record systems is costly, and requires farmers to be well educated and have a high learning capacity. The verification process itself, done by a trained, accredited auditor, is also expensive.

It is therefore not surprising that most standards are suitable primarily for farmers who are (already) better organized and educated. In practice, this means that sustainability standards are mostly implemented by those farmers who are already more sustainable (those at the "top part of the pyramid"). For the target group at the "bottom of the pyramid"—the unorganized, illiterate subsistence farmers—most standards prove far too difficult to implement, as well as being far too expensive.

And it is not always worth it

Another frequently heard argument is that standards require farmers to make all kinds of investments and changes, but the higher price the farmer finally receives for the certified product is not enough to cover these additional investments. This could mean that, in some cases, a certified farmer is economically worse off than an uncertified farmer.

Certifying poverty

Most traditional standards, such as Fairtrade, Rainforest Alliance, UTZ Certified, RSPO, RTRS, ASC, and many more, are highly focused on social and environmental practices. These have often been the founding principles and are the basis of their claim as a standard. However, they do not always pay attention to good business and management practices or to helping farmers run a proper, profitable business. This leads to the observation that many standards actually "certify poverty." Thus, the implementation of standards may have helped farmers to be slightly more socially and environmentally sustainable, but they nevertheless remain mired in poverty.

Certify or else ...

Finally, standards are often no longer a tool for the implementation of better practices, but are becoming a barrier to farmers who are not certified. If farmers do not have a certification, for example because it is too expensive or too difficult, they will be excluded from the market. This encourages farmers to sometimes have multiple certifications and, with that, multiple costs, because different clients demand different certification standards.

The combination of fast-growing demand for certified products and the difficulties inherent in implementing standards leads to a situation where farmers simply must become certified to continue selling their products. This, in some cases, may encourage fraud and false auditing. The brands think they are buying sustainable coffee or cocoa; the reality is that nothing has changed on the ground as the auditors have been bribed to pretend that the farmer was certified. This is not unimaginable in developing countries.

In short

Standards and certification programs are increasingly considered to be expensive tools that comprise only part of the solution. To really transform the market we need to move beyond the use of standards and address the other root causes of the problem.

Jan Kees Vis, former chair of the RSPO, reflected on the years that characterized the second phase of market transformation:

> I am relatively satisfied with what has been achieved so far. In recent years RSPO has delivered nearly everything it has promised, and 14% of world production has been certified in five years' time. No other voluntary sustainability standard has been able to achieve the same. But I think we could have done more and that we need to make some key decisions to ensure that we keep on this path.[259]

The story of the palm oil sector ends here ... for now. The sector has reached the end of the second phase of market transformation. To continue to be a driver for change, it will have to rethink its strategy. The palm oil sector can learn from how other sectors are currently doing so. My predictions are that the palm oil sector will move to a phase where the importance of standards will be less critical. More important will be to develop an holistic sector-wide vision on what true sustainability means for the sector, and how to get there; to identify the roles, responsibilities, and measures of success on how to get there, including the role of national governments; and to reorganize the sector in such a way that there is accountability about moving toward that objective. It will become very important that this process is locally organized and owned.

7.5 Other examples of agricultural sectors in the second phase of market transformation

What happened in coffee, cocoa, and palm oil has also happened by now in most agricultural sectors. There are many examples of agricultural sectors that are currently in the second phase of market transformation. And, as a result, dozens if not hundreds of standards emerged. What follows are short descriptions of the tea, livestock, soy, cut flower, aquaculture, sugarcane, cotton and tropical timber sectors.

7.5.1 The tea sector

Tea is produced mostly in Kenya, India, Sri Lanka, and China. In 2012, approximately 3.41 million tons of tea was produced worldwide. Tea production employs 8 million workers in China, 3 million workers in Kenya, 1.3 million workers in India, and 1 million workers in Sri Lanka.[260,261] The tea sector is currently in the second phase of market transformation, but is showing initial signs of moving toward a more collaborative approach (the third phase of market transformation).

What are the issues? The sector traditionally is plagued by sustainability issues such as poverty and poor working conditions (low wages, and low job and income security), child labor (especially on smallholder tea plantations), high use of pesticides and severe water pollution, biodiversity loss (tea is often cultivated in more rugged and remote areas, which tend to be areas with the highest biodiversity), soil erosion, and deforestation (cutting firewood from natural forests to dry the tea).

Where is the industry now? As we have seen in other sectors, the tea sector also has a rich history of competing on standards. The most well-known standards are Rainforest Alliance, UTZ Certified, Fairtrade, and Organic tea. Unilever played a particularly important role in bringing the tea sector into the second phase of market transformation when its large tea brands Lipton and PG Tips committed to becoming 100% Rainforest Alliance certified in 2007.[262] This move forced almost every Western tea brand to reconsider its position on sustainability. A complicating element is that the total amount of tea drunk in Western economies is relatively low (60% of the world tea production is consumed locally; the rest is exported to developing countries and Western markets),[263] so Western tea brands have limited leverage in terms of sourcing to change the dynamics in the tea sector.

A major step that signaled the end of the second phase of market transformation came in 2013 when the Indian tea sector established the "trustea" sustainable tea program. The main focus of the program is to accelerate the transformation of the Indian tea market in close partnership with tea industry stakeholders through the development and implementation of a tea sustainability code based on Indian realities. The India-specific tea sustainability code is an industry initiative, and will be developed by the industry for the industry.

7.5.2 The livestock sector

A global industry, the livestock sector has several serious issues threatening its future. Consequently, it is under growing scrutiny. The total world production of livestock in 2012 was more than 300 million tons, including 105 million tons of poultry.[264]

What are the main issues? The cattle industry causes severe water pollution from pollutants, animal wastes, antibiotics, hormones, chemicals from tanneries, and fertilizers and pesticides used for feed crops, which can create severe eutrophication and dead zones in seas and lakes. It is one of the largest sources of greenhouse gases, with livestock production accounting for 18% of global greenhouse gas emissions. Soil erosion is a serious problem: about 70% of all grazing land in dry areas is considered degraded, mostly due to overgrazing, compaction, and erosion attributable to livestock activity. Land use is extreme: the livestock sector is by far the largest user of land. It causes severe deforestation, especially in Latin America: 70% of previously forested land in the Amazon is used as pasture, and feed crops cover a large part of the remainder. Huge amounts of antibiotics are used: 50% of all antibiotics used worldwide are given to livestock. Finally, animal welfare is a major concern, with animals often raised under poor conditions.

Where is the sector now? Several initiatives and standards are addressing the poor practices in the sector. The main standards are: Animal Welfare Approved (AWA), Global Red Meat Standard (GRMS), GlobalGAP–Livestock, Food Safety System Certification 22000 (FSSC), Food Alliance, and EU Organic Farming. In 2013, the Sustainable Agriculture Initiative Platform (SAI) launched New Principles for Sustainable Beef Farming, endorsed by McDonald's Europe and Unilever. The sector is still very focused on standards, while the root problems continue.[265]

An interesting development in New Zealand is the Red Meat Primary Growth Partnership Collaboration Program. This is a seven-year program which brings together a number of participants in the red meat sector, including cooperatively owned and privately owned processing companies, which together account for a substantial majority of New Zealand's sheep and beef exports. This is one of the first signs that the sector is trying to reorganize itself regionally and address the issues collectively.

7.5.3 The soy sector

The soy sector is similar to the palm oil sector in many ways; they are both bulk commodities (hidden ingredients). Soy is mostly used as animal feed for raising livestock. Top producers of soy in 2012 were the U.S., Brazil, Argentina, China, and India. In that year, 212 million tons of soy was harvested from a production area of more than 90 million hectares (909,000 km^2), mostly on large mono-cropping farms.[266] The soy sector is currently in the second phase of market transformation and is trying to cope with some very serious issues.

What are the main issues? In many regions the soy sector is known for poor labor conditions and forced labor, violation of land rights, harmful agricultural practices, massive deforestation for cheap and quick expansion, degradation of ecologically sensitive areas, and soybean monoculture, which can cause soil erosion and nutrient depletion.

Where is the sector now? The sector is focused on the implementation of standards, including the Round Table on Responsible Soy (RTRS), GlobalGap, EU Organic Labels, Rainforest Alliance, ProTerra Certified, EcoSocial, IMO Fair for Life, and Fairtrade. The main standard is the RTRS, which is managed by the industry platform for responsible soy production.[267] Despite these various standards, the total amount of certified soy remains low, and deforestation continues on a large scale.

7.5.4 The cut flower sector

The cut flower sector is carefully taking its first steps toward collaboration beyond standards. But it is still at an early stage and insecure. Trade in flowers is generally regional because of the perishable nature of flowers. The most important producer and trader of flowers in Europe and, indeed, the world, is the Netherlands.[268] The second largest exporter of flowers in the world, Colombia, together with its neighboring country, Ecuador, supply the U.S. market. Israel (the third largest exporter of flowers in the world), India, Malaysia, and Thailand are important flower exporters in Asia.[269] The biggest East African floriculture industries are hosted in Ethiopia, Kenya, and Tanzania. Australia and New Zealand supply seasonal flowers that are impossible to cultivate in Europe and North America.[270]

What are the issues? Cut flowers are produced both in Western and developing countries. The issues therefore differ between these regions. In general, the sustainability issues are exploitative and informal labor conditions, discrimination, poor health and safety conditions, pollution, high water and or energy use, and high chemical usage.

Where is the industry now? The cut flower sector has more than 12 competing standards. A few examples of these are: Milieu Project Sierteelt (MPS), GlobalGAP, Rainforest Alliance, Fairtrade, Fair Flowers–Fair Plants, EHPEA (Ethiopian standard), and FlorVerde (Colombia).

In 2012, IDH initiated the Floriculture Sustainability Initiative (FSI) to create mainstream sustainability in the international floriculture sector and to introduce a benchmark standard to align the various competing standards. All members of FSI have to commit to raise the level of sustainability in their supply chain to 90% by 2020.

What are next steps? The FSI is a much-needed platform to enable this fragmented industry to become better connected and organized. Currently the focus is still very much on benchmarking the different standards and increase trust and collaboration. In time, the objective for the FSI should become to try move beyond the use of standards as instrument of change and collaborate on identifying the overarching issues, reaching out to governments and work more collectively.

7.5.5 The aquaculture (fish farming) sector

To supply our growing population, we can no longer rely on the seas alone to provide us with fish. Aquaculture, the farming of fish on land, is seen as the future for the fish supply industry, but only if it can be made more sustainable. China nowadays is responsible for 61.4% of total world production of farmed fish (36.7 million tons), followed by India (7.76%), Vietnam (4.46%), and Indonesia (3.85%).[271]

What are the issues? The sector is notorious for its negative environmental impact, particularly with regard to: release of organic wastes and toxic effluents into the oceans; overuse of antibiotics, pesticides, and disinfectants; destruction of coastal ecosystems; land seizures and displacement of coastal communities; depletion of freshwater sources to build aquaculture ponds; and escapes of cultured fish species into the wild. Moreover, fishmeal, which is made from wild fish, is used as a feed, which causes overfishing. Other problems include depletion and salinization of potable water and agricultural land, and lack of compliance with food safety standards.

Where is the sector now? The problems in the sector have attracted a lot of attention. In particular, the WWF has sounded the alarm bell for years and is actively trying to make the sector more sustainable. The sector has various standards, each with their own focus and background. The main standards are: Aquaculture Stewardship Council (ASC), Global Aquaculture Alliance (GAA), Friend of the Sea, GlobalGAP–Aquaculture, and EU Organic Farming.[272]

7.5.6 The sugarcane sector

Besides producing the common sugar used in food and beverages, thanks to high-tech innovations, the sugarcane sector has diversified and now produces ethanol, bioelectricity, bioplastics, and other products. This has led to an increased demand for sugarcane. However, the level of the sector's sustainability, and other complicated issues, raise doubts about its prosperity in the near future.[273]

What are the issues? Sugarcane is a thirsty crop. It uses water intensively and results in water scarcity. Due to run-off from applied fertilizers and industrial waste from sugar mills, water sources are being polluted. Sugarcane cultivation leads to loss of topsoil when grown on steep hillsides without terracing. What is more, sugarcane plantations are displacing food crops and occupying indigenous lands. Another important issue, and one that is attracting increasing attention, is the flouting of labor laws in the sector: extremely low wages, dangerous and harsh working conditions, and slave labor.

Where is the sector now? The main standards in the sugarcane sector are the Roundtable on Sustainable Biofuels and the multi-stakeholder certification platform Bonsucro, established in 2011. Various countries have separate initiatives, such as the Sustainable Sugarcane Initiative in India.

Big multinational companies are also starting sustainability programs in the sugarcane sector. Unilever and Solidaridad signed a Memorandum of Understanding to work together on sustainable sugarcane in Central America. Brazil, where sugarcane plantation expansion is a very sensitive issue, enacted a law banning the expansion of sugarcane plantations and the construction of new sugar and ethanol plants in three key fragile and high-biodiversity areas of Brazil—the Amazon, the Pantanal, and the Upper Paraguay River Basin.

7.5.7 The cotton sector

The popularity of "fast and cheap fashion" (the fast-changing fashion styles and cycles) has vastly increased the demand for cotton. However, the complexity of issues in the sector, and the after-effects of cotton cultivation, threaten the future of the sector. The biggest producer and consumer of cotton is China, which produces 27% and consumes 40.5% of the world's cotton. Other producers of cotton are India (21.8%), Pakistan (8.4%), and Brazil (7.3%), and other cotton-consuming countries are India (17.9%), the U.S. (12.7%), and Pakistan (9.5%).[274]

What are the issues? Cotton, like sugarcane, is a thirsty crop. Cotton production accounts for 3.5% of total worldwide water use and leads to water depletion in dry regions.[275] Cotton cultivation causes serious environmental pollution due to its use of chemicals and pesticides. Globally, cotton accounts for around 6.2% of pesticide sales, and up to 50% of all pesticides used in developing countries.[276] Low and volatile prices, and low productivity, cause poverty on smallholder cotton farms. The violation of labor laws is another sensitive issue in the cotton sector: child labor, forced labor, low wages, and poor and dangerous working conditions. A complicating factor in the cotton industry is the many different players and steps in the value chain, which makes it difficult to introduce traceability for sustainably grown cotton. This, in turn, makes it difficult to link sustainably grown cotton to the end products in the market and to make any credible market claims. The result is less competitive dynamics and leakage in the system, whereby investments in sustainability are not captured by the value chain.

Where is the sector now? The cotton sector is in the second phase of market transformation, with a number of certification systems: Organic cotton standards, Fairtrade, Cotton Made in Africa, and the sector roundtable, the Better Cotton Initiative (BCI), the main roundtable certification program. The issue of child labor has been highlighted many times in the international press. Separate individual initiatives were established, mainly by NGOs, with the WWF designing and implementing a field project in Pakistan and the Dutch NGO Solidaridad launching the MADE-BY label. Companies are also slowly getting involved; for example, IKEA funded a WWF project to introduce better farming practices.

7.5.8 The tropical timber sector

The timber sector shows the first signs of broader collaboration. However, laws to regulate illegal timber logging have not been enacted at international level, and timber-related policies in various countries or regions are not aligned. As a result, the issues in the sector have been mitigated slightly, but still persist.[277]

What are the issues? One of the biggest issues is illegal logging. Illegal logging has an extremely harmful effect on the timber market and the environment as well as impacting on tax revenues. It is believed that almost half of the timber in Indonesia, 25% in Russia, and 70% in Gabon originate from illegal logging. According to the World Bank, the annual market loss from illegal logging is U.S.$10 billion. In addition, governments lose U.S.$5 billion in revenues from taxes.[278] These illegal practices lead to environmental degradation from forest fires, biodiversity loss, and soil erosion. Furthermore, logging roads make forest lands accessible for further exploitation and forest plantations replace natural forests, which also has a negative impact on biodiversity.

Where is the industry now? The sector is currently stuck in the second phase for the time being. The main standards in the timber sector are: Forest Stewardship Council (FSC), Program for the Endorsement of Forest Certification (PEFC), and the Sustainable Forestry Initiative (SFI). Besides the standards, many national initiatives are combating illegal logging. An example of a larger-scale policy initiative is the EU FLEGT Action Plan 2003 and 2013 which counters illegal logging by providing support to timber-producing countries.

The EU and the U.S. have enacted laws that prohibit the marketing of illegal timber: the EU Timber Regulation (EUTR) and Lacey Act in the U.S. The next steps are clearly to move beyond illegal logging, to go beyond the discussion and competition between standards and seek a broader "industry-to-government" approach.

7.6 What is the next step for these sectors?

The sectors mentioned here are all in the second phase of market transformation. Some have been in this phase for more than ten years. They are all relying on sustainability standards as their main instrument of change. In all cases there are different standards competing with each other and, in all sectors, the issues are not getting solved. Pushing harder on standards will clearly not do the trick: a shift in gear is necessary.

The next step for these sectors is to transform to the third phase by moving beyond the use of standards to create an overarching vision that unites the industry to work together and aligns its efforts with national governments. Only then are the root causes of system failure addressed. The next chapters highlight examples of sectors that have already moved into the third phase of market transformation.

7.7 Reflections on the second phase of market transformation

The second phase of the market transformation curve is truly one of unseen market dynamics. This phase is all about fierce competition, mounting pressure to act, core business, strategic responses, and first-mover advantages. Most of the agricultural sectors are currently in this phase of development.

Driven by the insight that the problems in the sector will not go away by themselves, and that pressure from campaigns will continue, some companies are beginning to see the opportunities of becoming a first mover and claiming the high ground. If done correctly, the result can be predicted. A public commitment by one company forces other companies to reconsider their own position and act, especially if the action of the first mover is successful and they gain a first-mover advantage. Soon, companies do what companies do best, compete: but this time on sustainability. It is very important in this phase to ensure that first movers actually gain first-mover advantages. This can be in the form of additional brand value, NGO or media recognition, respect from peers or favorable clients, and a positive market response. The success of one company will force others to act, and this is the reaction we want.

However, companies cannot do this alone. This is the era of multi-stakeholder partnerships and programs. Companies might have the resources and the marketing power, but they lack local knowledge or presence. But, most importantly, they lack credibility. They need to find partners, and the standard organizations and leading NGOs offer the perfect match. Standards offer an instant, off-the-shelf solution: all the company has to do is commit to buying certified coffee, cocoa, or palm oil. The standards also give the company brand value and differentiation in the marketplace.

The golden rule of competition, however, is to imitate, but never follow. When first movers are successful, the second and third movers in the industry will try to take the advantage from the first movers by imitating them, and also by differentiating themselves from them, and adding value to their own brands. This competitive response is particularly powerful in sectors with branded products that compete in the consumer marketplace. Coffee and chocolate are examples of markets with branded consumer products where consumers understand what the ingredient is and where various brands compete for the favor of the consumers. In these markets it is natural for brands to want to differentiate themselves from their competitors. The situation is different in the case of "hidden" commodities such as palm oil, soy, or sugar, where consumers will not always know what the main ingredient is. Moreover, these ingredients are not part of the "brand story." Therefore, in these sectors, the competition does not focus on brand differentiation. Instead, competition between standards is driven much more by price, value chain efficiency, political power, and national interests. But, in both cases, competition on the standards and solutions is emerging.

In many cases, the result is the same. In a few years, various standards and dozens of programs emerge, and companies start competing on sustainability. As explained by Stefanie Miltenburg from Douwe Egberts Master Blenders:

> You had to differentiate to the outside world. Everyone had their own stories about "we invest so much, will train so many farmers and will do so much volume." It was really an arm wrestling match. We often looked at Kraft Foods (now Mondelēz), as we considered them to be our peer company. They were about the same size, and offered the same kinds of products. So when they announced a commitment to 20,000 tons of Rainforest Alliance certified, we would commit to about the same volume of UTZ Certified.... At that time it was really important to show everyone that we were doing at least as well as the others.[279]

Standards, in turn, thrive with competition, and the certifiers will work hard to ensure that multinationals choose their standard. This is important because demand for certified products triggers supply, and that drives the case for change. NGOs benefit from these partnerships as well because funds flow into their programs, and most NGOs enjoy the position of being recognized for their knowledge, and as the judge and jury about what is credible and what is not. It is a real win–win situation for all involved.

The standards also start competing with each other. According to Rob Cameron, formerly CEO of Fairtrade International:

> Companies would come to us and say: "We are thinking about putting an ethical label on our product, but we are not sure which to go with. We might look to UTZ, to Fairtrade, to Rainforest; come and talk to us and tell us what you can do and how you can help." Inevitably, you know you are in a competition because they can only put one label on, and if you believe your label is the best one, you are going to compete.[280]

Competition between standards is not necessarily a bad thing, as illustrated by Han de Groot, Executive Director of UTZ Certified: "I see competition between standards as a foot race. It's an individual sport, but if you run in a group, you go faster. You are grateful to your competitors because they make you better."[281]

I believe this is partly true. From experience I know that competition, the will to win, can lead to innovation, to faster change, and to a more dynamic standards landscape. That is why this second phase of market transformation is so important. However, the logic holds only up to a certain point and that is when the industry is no longer competing on impact and innovation, but instead gets locked into various standards, auditing, traceability, and labeling systems with unnecessarily high barriers to enter and barriers to switch between standards. That is when competition gets unhealthy and can actually become an impediment. This is when the movement toward more sustainable production is in danger of losing momentum altogether simply because the price is too high, it is not practical enough, and the added value is too small. In the meantime, the real problems in the sector are not being solved because they are more complex than just the implementation of standards.

According to Peter Erik Ywema, General Manager of the Sustainable Agricultural Initiative Platform:

> I have noticed an important development in recent years and that seems to be a game-changer, not necessarily in a positive way. Because there are so many standards, both consumers and professionals have lost their way. When we did our benchmark in 2009, about 25 standards were considered to be important. In the meantime, about 500 standards exist. Consequently, you don't have any idea about what to do for what market and what is right or wrong.[282]

More important is the notion that competing standards cannot change all the negative loops that are causing the system failure in the first place: failing markets, failing governance systems, and failing support systems. Standards are designed to help farmers implement better practices and help companies to source more responsibly. The lack of access to finance or access to services and inputs continues, and the lack of effective government support structures remains unaddressed. And as we have seen in Part I of this book, these are essential to complete the market transformation process.

In the meantime, the sustainability issues continue to affect the sector. They keep returning. Despite all the efforts, there is still extreme poverty among cocoa and coffee farmers, and the forests in Indonesia are still burning, even in 2014.[283] If competition on the implementation of standards is not working, then what does?

The insight is slowly dawning that these large companies are also part of an even bigger puzzle—the system itself—and the challenge is bigger than their direct spheres of influence. Therefore, what brought us *here* will not get us *there*. To solve the issues, the sector must attain higher levels of interconnectedness and "connectability." More and more companies are now starting to see that the future of their business is at stake. Without a vital and professional farmer base, there will not be a future for them either.

It is time to move to the next phase of market transformation: the critical mass and institutionalization phase.

Summary of the second phase of market transformation: the first mover and competition phase

Triggers for change	- Problems in the sector persist - Increasing insight that you can use sustainability to gain a competitive advantage
Main change agents	- First-mover companies - Standards organizations - In the case of cocoa, the U.S. Senate
Who is against change?	- Project owners of the first phase of market transformation - NGOs who resist working with the industry
Initial response and level of awareness	- The problem will not go away: we can and should do more - First movers: we can benefit by being the first to change - Sustainability is linked to our core business/product - Laggards: let's keep a low profile and hope attention for the topic goes away
Driving forces for the market	- Continued NGO campaigning and media pressure - Lawsuits - First-mover advantages: marketing and corporate social responsibility promotion - Pressure on laggards
Willingness to cooperate with others	- Growing willingness: others can be partners as long as they are not competitors
How to start this phase and be successful. What can you do?	- Create best practices/standards that give a competitive advantage to first movers - Give positive attention to first movers and put pressure on laggards - Spark competition
Limitations of solutions and barriers of this phase	- Farmer change is mainly driven by premiums, expensive certification programs, and NGO capacity-building support for farmers - Inefficient use of resources due to proliferation, fragmentation, and competition of standards - Programs can only reach certain number of farmers: not everyone will be included - At some point the added marketing value declines, while the costs of the programs continue to rise

What comes next?	Either: - Attention to sustainability decreases and the sector tends to fall back into its old behavior (until a new crisis happens) Or: - Companies start to realize that this is more than just a marketing program. Sustainability is starting to threaten the future viability of the sector - Slowly, the competing companies start to become used to the idea that a change in strategy is needed. To solve the issues in the sector, the rules of the game need to change, and this requires collaboration - The next phase of market transformation is up to the industry: the critical mass and institutionalization phase

8
The critical mass and institutionalization phase

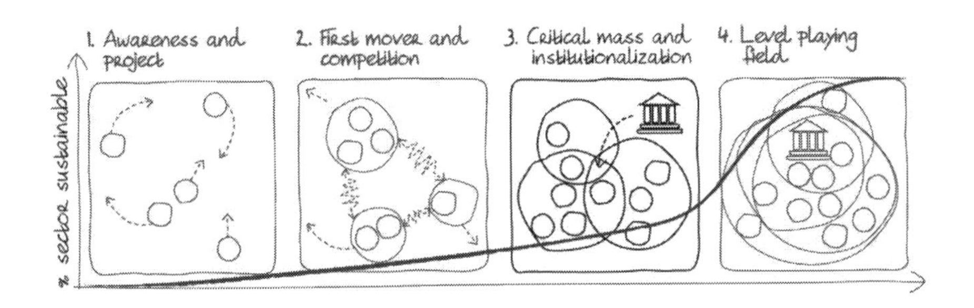

Important conclusions of this chapter:

1. The standards industry continues to grow, but is increasingly subjected to criticism. Why spend all that money on certification and premiums for farmers if it does not seem to structurally solve the problem? Moreover, the added value of certification seems to be diminishing.
2. In the meantime, concerns about the future of the sector are starting to grow. What will our sector be like in five to ten years' time if we do not act now?
3. Some strategically oriented companies see that it is time to change strategy. The fragmented, competitive approach may work well in the marketplace where companies need to differentiate. But behind the scenes some companies are starting to realize they need to work together and align efforts.

4. Then the critical mass and institutionalization phase starts. This is the most critical and most sensitive phase of all. Companies that otherwise are competing need to learn to trust each other, share information, and agree on high-level collaborative strategies. This includes working together with national governments, which in some cases seems to be an even bigger challenge from both sides.
5. While the leading companies come together on neutral ground, others feel excluded and threatened. This includes former allies such as the standard organizations and the NGOs who used to be companies' partners in change.
6. Gradually, acceptance of a high-level, collaborative strategy grows, and more and more companies and initiatives start to align themselves with it. If this phase is managed successfully, the entire sector is on the verge of a tipping point.

8.1 How the cocoa industry took leadership

"Participants in this meeting will respect anti-trust laws at all times. It is not permitted to enter into any discussions or agreements that may restrict competition or even be perceived as affecting competition in the marketplace. This includes any discussions on sensitive competitive information such as individual prices, production and sales forecasts, production capacities, costs, rates, market practices, investments, or any other competitive aspect of an individual company's operation ... Are there any questions?" The anti-trust lawyer looks around the room to see that everyone understands the seriousness of his admonition and agrees to abide by it.

It is January 2013, on a cold and windy day in Brussels, Belgium, and the streets are covered with snow. Inside the prestigious office of a well-respected international law firm, representatives of the world's seven largest cocoa brands and processors are meeting together: ADM, Barry Callebaut, Cargill, Hershey, Mars, Mondelēz, and Nestlé are holding a face-to-face meeting of their Technical Working Committee. The meeting was organized by the WCF, a neutral cocoa industry membership organization promoting a sustainable cocoa economy, as part of an unprecedented senior management commitment made a few months earlier to work side by side to take on a challenge that threatens the future of the sector. The meeting is facilitated by consultants and monitored carefully by anti-trust lawyers in accordance with anti-trust compliance guidelines.

The atmosphere in the room is initially rather tense. Polite words, small talk, and introductions are exchanged. Most of the people in the room know each other. They have seen each other during conferences, or worked together in working groups and committees. But today is different. Today they have come together to discuss the question that all of them now consider crucial for the survival and viability of the cocoa sector: "How can we, as an industry, help to restore cocoa farming as a sustainable and attractive profession for farmers?"

None of the companies who are represented in the room are strangers to the topic of sustainability. In fact, all of them have invested considerable resources, time, and effort in trying to deal with the issues in the cocoa sector. All these companies have been members of the WCF for a long time, and have contributed to many cocoa farmer support programs. Each of the companies has independently committed to buying increasing amounts of cocoa that has been certified by one of the leading sustainability standards, such as Rainforest Alliance, UTZ Certified, or Fairtrade. Almost all of them have launched projects with compelling names such as Cocoa Horizons, Cocoa Life Program, CocoaLink, Cocoa Livelihoods Program, Sustainable Cocoa Initiative, PACTS, Cocoa Promise, and the Cocoa Quality & Productivity Program.[284] And all of them became partners in one of the many large donor, NGO, or government-driven programs.

In each of these programs farmers were trained, schools were built, communities strengthened, certification standards implemented, productivity increased, and quality improved. Hundreds and hundreds of millions of dollars were spent on programs to help cocoa farmers. Nevertheless, cocoa continues to be "the poor man's crop." Almost all cocoa farming households in West Africa live well below the absolute poverty line, and working conditions are harsh. Every day the cocoa farmers and their cocoa trees grow older. Today, the average age of a cocoa farmer in West Africa is between 55 and 60 years old. The average life expectancy is about 65 years. This means that, in the next five to seven years, the industry expects to lose more than half of its farming population due to old age.

To make matters worse, it is becoming painfully clear that the children of these farmers do not want to take over their parents' profession. They have seen the backbreaking work, the struggles and the hand-to-mouth lives of their parents, and they do not want that for their own future. Instead, they choose to become a palm oil or rubber farmer, or to sell their land to gold miners and try their luck elsewhere, perhaps in the city. Anything is better than being a cocoa farmer. Everyone realizes that when younger generations do not want to take over the farms, it is only a matter of time before the sector starts to die.

This is one of the reasons cocoa shortages are predicted. Recently, Goldman Sachs estimated that the 2012–13 crop seasons would have a shortfall of 100,000 metric tons, and some companies are expecting this shortfall to increase to 1 million metric tons of cocoa by 2020, particularly because of increasing expected demand from countries such as China and India.[285] As a result, cocoa prices are expected to rise and become even more volatile than they already are. However, the higher prices do not seem to trickle down sufficiently to the farmers and do not appear to motivate them to improve their farming practices. On the contrary, the farmers, their trees, and soils are becoming less and less productive.

It is this gloomy foresight that brought the seven leading cocoa companies together on this cold day in Brussels. After a presentation by the consultants, who analyzed the effectiveness and impact of the many sustainability programs to date, participants were asked to envision a thriving and sustainable cocoa sector. Even though they are competing companies, it didn't take long for them to articulate an overarching compelling vision that unites them in their objectives and actions. This vision was the idea that cocoa farming must again become a profession of which farmers can be proud. Farmers need to make enough money to send their children to school, feed their families, and invest in their farms. We must stop looking at cocoa farmers as poor people in need of help. Instead cocoa farming should again become a profitable and attractive business of choice! This is what the future should look like, and all the participants understand that we have less than a decade to achieve this aim.

While this vision may sound simple and even logical to some, its realization will be nothing less than a Herculean task. To increase farm income to compete with other sectors and provide for the daily needs of their families, the cocoa farmers must become much more productive and efficient. To increase productivity, the farmers need to be more professional in their farming techniques, plant younger and more productive trees, and enrich their soils. To do this on the required scale means training hundreds of thousands of farmers, developing a non-existent fertilizer industry that can start providing inputs to the farmers, and setting up massive replanting programs and facilities, which are in many cases non-existent as well. And this is just the agricultural side of things.

Realizing the vision also means building thriving cocoa-growing communities: places where people want to live and build their livelihoods. It means empowered communities that can articulate their own needs, take matters into their own hands, and stand up for their rights. It means kids going to school instead of working in the fields. At school they learn either how to become good, professional cocoa farmers, or how to find another respectable profession. This, in turn, requires empowering female farmers and giving them an equal say in their communities and households, especially on how farm income is spent; after all, they do most of the work in the fields and are often the backbones of households and farms.

It essentially requires reorganizing and rebuilding cocoa farming in West Africa as well as in many other parts of the world. As you can imagine, this challenge is too big for a single company, and the companies now understand this. It is even too big for the consortium of the seven largest cocoa companies who joined forces in Brussels that day. Although these companies initiated and led this strategy, they agreed that they could not accomplish it by themselves. Ultimately, large parts of the cocoa industry need to be aligned with this vision, share the same notion of urgency, accept its implications, and start working together. Even more importantly, the overarching vision should be coordinated with national governments, ministries, and institutes. These public agencies are tremendously important to the realization of the vision. The same is true for large donors such as the World Bank, development agencies and charity organizations, funding agencies, and sustainability standards. The first challenge is to stop the fragmentation of all of these programs and projects, and to align the interventions and resources instead around one common vision and one set of key performance indicators (KPIs). Or, in the words of one of the participants:

> The fact that we are competing in the marketplace, and will continue to do so, does not mean we cannot define common objectives together and agree on similar strategies, use the same KPIs and share what we learn, all designed and intended to help farmers enjoy better living conditions. We cannot wait any longer for others to help us out. It is the cocoa industry that needs to take the lead, because this is our business and our value chains.[286]

As summarized by Alastair Child, Cocoa Sustainability Director at Mars:

> Our aspiration is that we can create a common set of priorities and KPIs from an industry perspective that are good for the industry, good for the farmer, and ultimately benefit the consumer. By aligning the efforts of many behind a core few game-changing interventions, such as training and inputs for farmers that lead to adoption of modern high-yielding farming techniques, farmers will benefit significantly—tripling yields and doubling incomes from current levels in West Africa—and industry will in turn benefit from long-term security of supply. This mutuality of benefit creates a compelling return for both, and this shared benefit is the key for enduring change.[287]

The seven companies of the Technical Working Committee continued to meet frequently to work on the vision. The level of trust increased rapidly between the parties as they more openly discussed the issues that prevent their sector from changing. This didn't mean they all agreed with each other at all times. But each company realized that it was important to listen and make agreements, while the presence of anti-trust lawyers ensured that everything that was shared, discussed, or agreed upon stayed away from competitive issues and within the boundaries of anti-trust laws.

Bill Guyton, President of the WCF and organizer of this new strategy, described this approach:

> We look to build consensus, and sometimes that is not easy on topics where companies feel differently, also because they are competitors in the marketplace. One of the things that unites

> the companies is that we are trying to put the farmer first. The focus on how interventions are helping farmers is of common interest to everyone. That is the main non-competitive issue we focus on.[288]

During the subsequent year, the group was able to work out the overarching strategy in much more detail. The time came to increase the number of participants of the industry. Five other large cocoa companies joined the initiative: Armajaro, Blommer, Ecom, Ferrero, and Olam. Now, the 12 largest chocolate and cocoa brands in the entire industry were at the same table. The name of the strategic platform was changed to CocoaAction. To ensure that CocoaAction would eventually become the strategy for the whole cocoa industry, the senior executives of the 12 cocoa companies also became the daily board of directors of the WCF itself.

As this book went to press (November 2014), CocoaAction was in its final phase of development and almost ready for implementation. The first announcements and high-level visits to national governments were taking place.[289] CocoaAction is expected to mark the beginning of a major, structural change in the cocoa industry.

CocoaAction represents a great example of an industry that has moved into the third phase of market transformation and is taking its destiny into its own hands. The sector has learned that complex challenges threatening the industry can only be solved if a critical mass of stakeholders can overcome their own short-term interests and work together on an overarching, comprehensive, long-term agenda. The 12 cocoa companies deserve recognition for their leadership and commitment. Never before has a group of high-level senior executives and cocoa industry leaders joined forces and made strong commitments to tackle their sector's structural issues. They are overcoming their immediate competitive reflexes to benefit a higher cause, to realize a higher vision, and that is admirable.

This is the third phase of market transformation, the phase of critical mass and institutionalization. It requires non-competitive, strategic collaboration that goes beyond the use of standards, and will very probably be the future for many other sectors that are still stuck in the second phase of market transformation.

8.2 Collaborative action in the coffee sector

It is clear that the coffee sector is under pressure. Leif Pedersen, Senior Commodities Advisor at UNDP, described the current situation in many coffee-producing countries as follows:

> Colombia and countries in Central America indicate that there is not enough labor available for the coffee harvest. Some farmers have to let their crop just drop to the ground because they do not have enough people to harvest. They are in competition with the general economic development, including the industrialization, of these developing countries, where people now have other choices. Prices for labor are going up and if the farmers do not get a better price for their coffee, their farms quickly become unprofitable.[290]

In 2012, the world's three largest coffee roasters, Nestlé, Mondelēz, and Douwe Egberts, together with the Tchibo (the fourth largest roaster in the EU), took the threat to their industry seriously and came together. They met on neutral ground at the office of IDH, the Dutch sustainable trade initiative, together with other parties, such as KNVKT (The Royal Dutch association of coffee roasters and tea packers), ECF (European Coffee Foundation), GIZ (German institute for foreign development cooperation), and Oxfam Novib as a representative for a consortium of other NGOs. The question on everybody's mind was: What does the future of the coffee sector look like, and how can we make it sustainable?

The people sitting at the table were the same people who led the market transformation curve in the second phase years before. There was a time when the coffee sector led the movement toward sustainability. The peak years were 2008–10 when the sector dominated the media with public claims of competing coffee companies committing themselves to sourcing increasing amounts of certified coffee, with the Dutch coffee sector signing a memorandum of understanding with the Dutch government stating that 75% of all coffee would be certified by 2015. Since then, however, not much had happened. Sure, the amount of certified coffee bought each year increased, but the upward movement, the energy and momentum, seem to have become lost.

The coffee sector was no longer a sector in transformation. After years of competitive dynamics, the media, donors—and even the NGOs—had lost interest in the coffee sector. Many believed that the issue was solved when the industry made a commitment to buy certified coffee. Coffee was on its way to sustainability; the problem was solved!

But the sector was far from sustainable and the problems were far from solved. Poverty was still prevalent. Productivity was still low and falling, the sector was still struggling with crop disease and had proved to be very sensitive to the effects of climate change.[291,292] The reality was that, after more than a decade of implementing standards and certifications, only 15% of coffee produced in 2012 was considered to be sustainable.[293] Considering the starting point, this is an impressive result, but it also means that 85% of coffee is not certified, and therefore probably not sustainable. And for the portion of the market that was certified, the impact was hard to prove.

As in many other industries, the criticism of standards as instruments for change was on the increase. The organizations responsible for the standards (the certifiers), who once sparked the competitive drive, needed to prove their added value and impact. Many reports and studies were commissioned to demonstrate the value of certification, such as *From Bean to Cup: How Consumer Choice Impacts Upon Coffee Producers and the Environment* (2005), *The Impact of Fair Trade Labelling on Small-scale Producers* (2009), *Toward Sustainability: The Roles and Limitations of Certification* (2012), *The COSA Measuring Sustainability Report: Coffee and Cocoa in 12 countries* (2014), and *The State of Sustainability Initiatives Review 2014: Standards and the Green Economy* (2014).[294,295,296,297,298]

The case for change and the evidence for impact were mixed, at best. The standards did not seem to have the transformational effects that most people thought or hoped they would. Organizations started to understand they needed to do more than just implement standards and place labels on packaging. Stefanie Miltenburg from Douwe Egberts also realized this:

> I believe that the certification programs we all endorsed and implemented over the past years did not bring all the change that was needed. It did help to connect the value chain and start the change, and it definitely helped certain groups of

> farmers. But many farmers did not benefit. Certification has not proven to be the solution to all problems. It is not enough to realize full mainstream change.[299]

It was this realization that brought the four leading coffee companies and other sector representatives together. Their aim was to rekindle the desire for change that once burned so fiercely in the sector and to bring parties together to start working on real solutions.

8.2.1 Meeting on neutral grounds

Competitors who want to work together need neutral facilitators. For the coffee sector, IDH (the sustainable trade initiative) proved to be a great facilitator. IDH is specifically mandated by the Dutch government to work with the private sector to accelerate and upscale sustainability initiatives within mainstream commodity markets and supply chains.[300] According to Joost Oorthuizen, Executive Director at IDH:

> We see that large companies which have been involved in sustainability initiatives so far have a real need for what we call "neutral but committed conveners." Companies need to work together, but as competitors they find it hard to do so. For this they need an organization that is committed, understands the business, is credible, and mandated to do so.[301]

Because IDH had been in operation for several years, it had already successfully adopted and initiated sector programs in cocoa, cotton, soy, spices, floriculture, and aquaculture. The coffee sector was to become the next success story. Joost Oorthuizen described their strategy:

> In the coffee sector we have been really successful in bringing together the four leading coffee roasters, and helping them formulate a non-competitive agenda. This was the first time in tropical commodities that something like this happened. The agenda was a long-term strategy, focused on the six most important coffee producing countries, and an aligned policy, investment, and outreach strategy. To this day this is unique in the coffee sector.[302]

Behind the scenes, and away from the critical eyes and voices of NGOs and standards organizations, the four companies took a look in the mirror and articulated their own vision of a sustainable coffee sector. According to the roasters, a sustainable coffee sector is one in which resilient and adaptable coffee farmers are able to successfully meet the needs of a constantly changing industry. It means that coffee farmers have attained a significant increase in productivity, that they have learned climate-smart coffee growing practices, and that they have access to finance.

Facilitated and guided by IDH and consultants, the program quickly took shape. The 4C organization, with its large membership base in the coffee sector, was also invited to the table. Together, an overarching strategy was formulated. This started to look like an initiative that could really change how business was being done.

Stefanie Miltenburg, who attended these initial meetings, observed:

> I think it was a hopeful sign that the industry was able to formulate its own vision of what a sustainable coffee sector looks like. But there is still a lot that needs to be done, which will require that we—as roasters—all look beyond and are willing to invest beyond our own agenda.[303]

What began in 2012, with four roasters meeting on neutral ground at IDH, has gained momentum and has started to spread. Very recently, the 4C initiative held its own special session to formulate Vision 2020 for the coffee sector. The number of participants for this event exceeded expectations. Over 40 representatives from trade and industry, coffee-producing countries, and civil society joined together, not to discuss the development of standards, but to define a common vision for a sustainable coffee sector as a whole. Cornel Kuhrt, a coffee expert with over 20 years of coffee industry experience, explained:

> The energy is coming back to the sector, we are moving ahead. The activities we see now in the sector are the Sustainable Coffee Program powered by IDH and Vision 2020 of the Coffee Sector initiated by 4C. These activities address big questions for a more sustainable coffee sector such as: Where do we want to go as a sector? How to organize non-competitive activities in the coffee sector in a more efficient way and how to integrate all the stakeholders we need? What are the needs of the coffee sector to improve toward a more sustainable sector? What to focus on and how to become more farmer-centric? And how can we address various issues pressing beyond certification standards? ... We are not there yet, but we have identified the next topics we have to address together.[304]

The sector is clearly in a process of reorganizing itself toward a higher level of interconnectedness and "connected ability" to be able to address the more complex problems that are threatening its existence. The sector is now joining forces behind a more overarching and more compelling vision that goes beyond the competition on sustainability programs and standards. The coffee sector has entered the third phase of market transformation.

For the next few years the sector is expected to remain in this phase. It will take time before the overarching vision of what sustainability means for the coffee sector has been translated into practical and actionable programs and measures. And it will take time before a critical mass of stakeholders from both the industry and the national governments have rallied behind it.

8.3 Reflections on the third phase of market transformation

According to Carsten Schmitz-Hoffmann, former Senior Executive at the 4C standard and currently Executive Director Private Sector Cooperation at GIZ, the German institute for foreign development cooperation: "The industry is shifting away from rigorous certification systems, and is getting skeptical about the additional costs of certification. Instead it is becoming more and more interested in stimulating intrinsic change."[305]

The third phase of market transformation is fundamentally different than the previous phases. In the cocoa sector, 12 of the largest cocoa companies, all competitors, joined forces to develop a shared vision about what sustainability means to their sector. In the coffee sector, four of the biggest coffee roasters joined forces to redefine what sustainability means for them, and later this grew into a movement of 40 companies. This realignment in thinking was unthinkable in the previous phases of market transformation. Sustainability evolved from a non-issue in the awareness and project phase, to a competitive issue in the first movers phase. In the third phase it becomes a non-competitive and collaborative issue and a business continuity concern. In this phase, competitors start to realize: "If we still want to exist as a sector in five to ten years, we really need to take sustainability seriously," and "This problem is way too big for us alone; we need to work together with competitors and governments to really change the way we work." Compare this to the reactions of the industry in the first phase of market transformation: we truly have come a long way.

It has taken most industries over ten years to evolve to this level of awareness, this level of trust, this level of interconnectedness, to arrive at this phase. Gradually, it starts to feel obvious for companies that they need to work together to solve bigger issues. In the words of Peter Erik Ywema, General Manager of the Sustainable Agricultural Initiative Platform:

> Someone told me a couple of years ago: "If you want to go fast, go alone. But if you want to go far, go together." The latter is more important in my business. It doesn't make sense to jump on the bandwagon of the fastest initiative. The art of my profession is to keep everyone together—even if that means that we have to go slower.[306]

But bringing competitors together, and having them work together toward a common objective, is easier said than done. It sometimes feels like magnetic repulsion: their natural instincts are to push each other away. Therefore, it is very important that a neutral, committed convener and facilitator holds the opponents together and facilitates the trust process. This convener needs to make sure that between competitors there is no favoritism, that information is shared equally, that roles and responsibilities are clear, that people are accountable, and that anti-trust rules are obeyed at all times.

The mandate of the convener is often structured in a governance system in which various stakeholders jointly take ownership and accountability for their collective vision and actions. Lise Melvin, former Director of the BCI, explains the power of a clear and effective vision for the sector:

> Getting a shared vision right and spending enough time on it is absolutely essential. Because these systems are so complex, it is very easy to become obsessed with one part of it. It helps you to ask: "Does what we do help us to realize our overarching vision" rather than ending up going round and round in circles, and being led by vested interests.[307]

In most agricultural sectors, the new vision that is formulated by the industry no longer refers to farmers who are poor and need help. Almost everyone foresees or wants a future in which the farmers are productive, resilient, efficient, and professional, in which they are farmers by choice and are proud of what they do—a future in which farming is increasingly seen as both a business and an opportunity. This is a far cry from the language that was used in earlier phases of market transformation, which focused on poverty, child labor, and environmental degradation, and farmers were essentially a problem to be solved.

This immediately raises the question of what will happen to farmers who are not business-minded, or who lack the basic education to become successful entrepreneurs. I strongly believe that answering this question will ultimately drive real transformation in agriculture. We must accept the idea that farming should no longer mean subsistence farming, but a business that generates sufficient food and income to support a prosperous life. This will have consequences for who can become a farmer and who cannot, and for how large farms should be to be able to generate sufficient income. This means that for most of the current subsistence farmers we will have to find alternative jobs or ways to generate an income other than relying on a plot of land to grow something to eat.

This also has consequences for how we view development aid to farmers. The collective sustainability programs, charity, and investments should go to farmers who ultimately can become independent from this aid. According to Annemieke Wijn: "The supply chain needs to pay for the initial investments, but I firmly believe that sustainable production is also more efficient and effective production. So once farmers are on their way toward sustainable production, they should be able to pay additional farm investment out of increased profits."[308]

Typical for the third phase of market transformation is the discussion on how to collectively create a landscape that enables and favors the more commercially oriented farmers. For the better farmers, inputs such as fertilizer, seedlings, knowledge, training, and access to finance should be available. Indeed, this is essential. At the same time, the issues of treating farming as a business, and increasing productivity, cannot be decoupled from basic issues such as education, child labor, and community-related matters. The reasoning is simple. Before the next generation of more commercial farmers can take over, they need to be educated and trained. First and foremost, this requires healthy and well-fed children to be in school learning how to read, write, and think analytically, not breaking their backs in the fields. And for the freshly educated children to become farmers and not move to the city to find other careers, they have to grow up in a vital rural community where they actually want to live. These matters go hand in hand. The future of sustainable agricultural sectors is therefore equally dependent on having educated people who choose to become farmers as well as on having thriving rural communities in which farmers can raise their families.

8.3.1 Resisting change

Some organizations or individuals always resist change toward the next phase. Mostly, these organizations have a vested interest in the status quo. This means that, in some cases, resistance may come from places you would least expect. Standards organizations, for example, the main drivers for change in the second phase of market transformation, do not necessarily favor strategies in which the industry connects and makes its own definition of sustainability without them. NGOs might also resist the structural change in this phase, as they might feel they are no longer needed; or that their role as endorsers or equal partners is diminishing. In many cases in this phase we see that the campaign-oriented NGOs start doing again what they do best: they start to launch campaigns on new issues. The transformation curve is again set in motion. The latest behind-the-brand campaign from Oxfam is a good example of a campaign-oriented NGO that has decided that its role is to once again shape the agenda for the industry, this time on matters such as land use, transparency, women's rights, and water use.[309]

More important is gaining support and cooperation from national governments. This is a critical issue to take into account. This is absolutely crucial for designing a landscape in which the more commercial farmers are able to thrive and alternative livelihoods are provided for the non-commercial farmers. However, some national governments may not be comfortable with industries uniting and aligning on a common vision. They may feel that they have not been properly involved in the process, which is often a valid point; after all, it is their industry too. Another reason for their hesitancy could be that an aligned industry is less easy to negotiate with and requires them to step up and become equal partners in the change process. It may reveal their true capabilities and interests in becoming an equal partner and committing to a sustainable sector. Restoring trust between industry and government and agreeing on a common long-term objective is a long and sometimes painstaking process, but critical for a sustainable future of any sector.

Resistance from other companies in the industry is also to be expected. Not all companies in the industry will have the same level of awareness or will understand the problem in the same way; perhaps they will intellectually still be at a level of thinking from a previous phase. Other companies may even want to exploit their position as laggards to gain competitive cost advantages over others who are investing and committing. This is a threat that, if not addressed, may again trigger competitive forces and the corresponding race to the bottom, which could negatively impact the level of collaboration within the industry.

All in all, this phase requires some time and patience to build trust between all parties and get aligned; but when enough of the sector becomes aligned with the overarching strategy, the process of market transformation speeds up. The sector is then able to collectively address overarching issues and resets the rules of the game to reward the right behavior and move to a higher connectability. If this happens, the critical mass phase is successful.

To push laggards, and to prevent a fallback into competition, the leading companies will at some point start to push for institutionalization, regulation, and even legislation. It might seem strange for companies to be pushing for more regulation, but the reality is that we see this all the time. If a leading majority develops an interest in stricter regulation, then they will lobby to enforce it. It forces a higher standard on their competitors, it levels the playing field, and it removes false competition. For the leading companies this makes sense, because they have already made their investments and gained first-mover competitive advantage, and are already complying with the new standard. The latecomers, on the other hand, will have no choice but to follow once regulation or legislation is in place. This means following the others, and accepting the costs as well;

but this time there are fewer rewards—it is just compliance. In this way, the playing field becomes level for *every* company. Sustainable behavior will then become a qualifier—if you don't comply, you are no longer qualified to be in business. And this is when we enter the next phase of market transformation: the level playing field.

8.3.2 Other examples of agricultural sectors in this phase of market transformation

Most of the agricultural sectors are currently in the second phase of market transformation. Besides cocoa and coffee, there no examples of sectors (to my knowledge) that have reached the third phase where an industry is collectively building its own vision of sustainability and collectively working towards this. However, several sectors are moving towards this transition point.

8.4 Summary of the third phase

Sectors like coffee and cocoa have entered the third phase of market transformation. A lot has been achieved. Competitors came together, aligned their visions and agendas, and started working on common goals, not just within the industry but also with national governments. The definition of sustainability is shifting to a more business-driven agenda as NGOs usually play less of an influential role in this phase. Some become implementing partners; others fall back in their campaigning role to push the industry forward. Slowly, the sectors are becoming ready for structural change. But it is a long and sensitive process. Trust and consensus between industry partners and between industry and governments needs to be built, and this takes time and many learning curves.

Some organizations will continue to lag behind and resist change, or refuse to contribute to the change and try to free-ride the process. To prevent a pullback, at some time lobbying will start to influence legislation to force the laggards to comply. This will eventually initiate the fourth level of market transformation: the level playing field!

Summary of this phase of market transformation: the critical mass and institutionalization phase

Triggers for change	- Problems persist and companies realize that the problem is bigger than competition can solve by itself - Awareness is increasing that sustainability issues are starting to threaten the future vitality of the sector
Main change agents	- Neutral convening platforms and industry representative groups - Leading industry groups in which former competitors work together - Governments may follow and provide support
Who is against change?	- Resistance may come from winners of the previous phase (standards organizations, NGOs, lagging companies) - National governments may resist change because they feel they have not been involved and are required to step up and commit
Initial response and level of awareness	- Awareness is high: the problem is so severe that it will be bad for business in the long run as business continuity is at stake - The industry and national governments need to work together, invest, and change the rules of the game. Impact and change on the ground is key
Driving force for the market	- Longer-term vitality of the sector - Securing sustainable sourcing - Efficiency of sustainability effort
Willingness to cooperate with others	- High, but suspicious in the beginning: companies can only do this if they work together, but they are still competitors in the marketplace
How to start this phase and be successful	- Bring industry leaders together: create an overarching vision and create clear transition pathways, joint KPI frameworks, and roles and accountability - Make the new strategy more inclusive - Hold players accountable and put pressure on the laggards

Limitations of solutions and barriers of this phase	- Lack of trust between the parties to collaborate and share knowledge - Lack of clarity about where the industry works together and where it competes
What comes next?	- Organizations start lobbying for policies to make unsustainable sourcing practices illegal or for policies that reward more sustainable practices

9
The level playing field phase

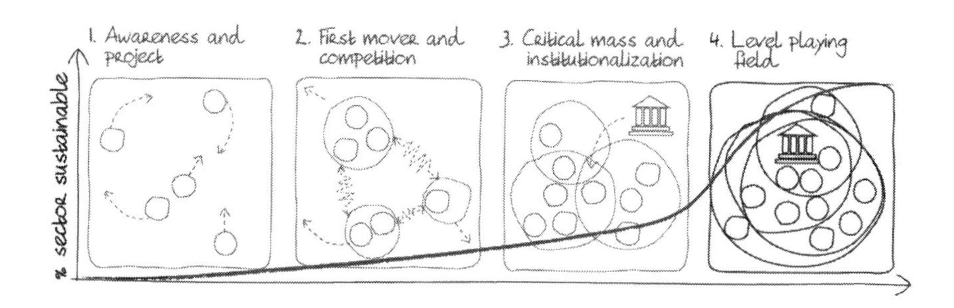

Important conclusions of this chapter:

None of the described sector cases in this book is currently in this fourth phase. This chapter therefore looks at other sectors (some non-agricultural) that have already entered the fourth phase and that can serve as an example of what can be expected for the commodity sector cases addressed in this book. Based on these examples and the forecast of the transformation curve, a future scenario will be described of what might unfold in the agricultural sectors we have been following up till now.

1. **After a successful period of industry alignment and collaboration, the sector is ready to move to the next and final phase of the market transformation curve.**
2. **Overcoming resistance and free-riding from laggards becomes a strategic issue. Policy barriers also need to be removed.**

3. An active international lobby starts to institutionalize aspects that large parts of the industry are already implementing, and make them obligatory.
4. Governments from both sides of the value chain are seeking cooperation, and their strategies are aligned with those of the industry.
5. As the synergies and the economies of scale start to take effect, and increasing numbers of farmers benefit from the changing systemic conditions, governments can claim their final and most important role: codifying proven practices.
6. Sustainability, a topic that in the previous phases has been the subject of campaigns, countless programs, competition, and sector alignment, is about to become the industry norm—a qualifier for doing business.
7. In the meantime, new issues have been identified and addressed. The transformation curve is starting once again.

9.1 Learning from other examples that have reached the fourth phase

Of the sectors we have been following throughout this book (cocoa, coffee, and palm oil), none has yet reached the fourth phase. In fact, this phase may not occur until well into the future. The cocoa and coffee sectors are examples of probably the most mature sectors in which the industry has come a long way. In these sectors we have seen an evolution of where sustainability has progressed from being a non-issue, to a competitive issue, to where the industry is working together and even reaching out to national governments to find ways to jointly create an environment where the sector can flourish once again.

When a critical mass of the industry has reached this level of awareness, of connectability and of alignment, it is time to *level the playing field* and formalize, institutionalize, and codify the new norms. We don't know yet what shape this will take for the sectors we have been following as we do not have examples to learn from. However, we can look at other sectors or products that have reached this level already and see what a level-playing-field phase has meant for them.

Three examples of sectors/products serve this purpose: the egg industry (also an agricultural sector), the mobile phone industry/phone chargers, and the lighting industry/compact fluorescent lamps. All three have reached the phase where Western governments have made sustainability the new norm. And although these sectors differ in many ways, I believe, their similar patterns of evolvement can help to foreshadow the future of sustainable agriculture.

9.1.1 Banning factory eggs

For years the EU egg sector has been under massive scrutiny from animal welfare campaigners. The living conditions of chickens in battery cages were appalling. Several decades ago, the egg sector became industrialized so it could compete on costs with other producing countries and comply with increasingly stringent food safety standards at the same time. But this came at the cost of animal welfare. The first attempts to change this situation were simple projects promoting "farm eggs": buy-from-the farm concepts. It did not take long before the first standards and labels appeared in the supermarkets, signaling the start of the second phase of market transformation. The number of standards and product claims grew significantly in this phase and included organic, free-range, humane, animal welfare approved, cage-free, free-roaming, corn-fed ... the list could go on. For years, the various standards and labels dominated the supermarket shelves. And in all this time, animal welfare campaign groups continued to pressure industry to find structural solutions to the animal welfare problem.

In 1999, after years of lobbying by protest groups and parts of the industry, and under increasing political pressure, the EU passed Directive 1999/74/EC banning battery cages in the EU from January 1, 2012. Germany, Austria, the Netherlands, and Sweden were able to eliminate all battery cages before 2012. Not all countries moved as fast. Six countries, including Portugal, Poland, and Romania, admitted they would not be ready in 2012, while Spain and Italy, among others, did not know whether they would meet the deadline. This led to fears that cheaper, illegal eggs, particularly liquid egg products from non-compliant states, would flood the market and undercut compliant egg producers.

As of January 2012, 5% of egg producers were still not compliant, and their licenses were revoked. In France, the damage was greater, as one-third of egg producers went out of business, according to figures from the industry association.

This example shows the effort and time it takes for industry and government to create a level-playing-field situation and to make new standards the new norm—even in a relatively well-controlled market like the EU.

9.1.2 Banning useless electronic waste

Another example of a sector that reached the fourth phase of market transformation is the banning of differentiated mobile phone chargers. This didn't come through competition on sustainability or voluntary compliance; it was forced upon the industry. In the past, every brand designed and manufactured its own unique mobile phone charger. This allowed the industry to have rapid innovations in how phones were charged and it enabled brand differentiation, consumer lock-in, and increased sales of product accessories. This served the industry well and for years it did not feel any urgency to change. For the consumer, however, it was impossible to use the same charger for a different brand or even a different product from the same brand. At one point, more than 30 different chargers were on the market, and almost all of us have drawers filled with such useless devices. This situation was both impractical and expensive for the consumer, and it also led to an incredible amount of waste: over 51,000 tons of electric waste per year in the EU from discarded chargers (obsolete, but still working perfectly).[310]

Due to consumer complaints, the European Commission decided to address the issue, as the industry itself had not taken any serious action. In March 2009, the Commission gave phone manufacturers an ultimatum: either adopt a common charger voluntarily or be subject to mandatory EU legislation. This ultimatum resulted in Europe's major mobile phone manufacturers agreeing to adopt a common charger for all data-enabled mobile phones sold in the EU. In 2011, European consumers were able to purchase a standard mobile phone charger. Most of the major players now use the same micro-USB as a charger interface. Standardizing the chargers and making this mandatory has ultimately benefited both consumers and the environment by forcing the industry to move into the level-playing-field phase.[311]

9.1.3 Switching off the lights

A third example of a sector that has reached the fourth phase of market transformation is the market for energy-wasting light bulbs.[312] Here as well, the previous phases of the transformation curve can be recognized, although governments were once again the driver of the transformation in most cases.

In the late 1980s, the first campaigns started in Denmark to raise consumer awareness about the high energy consumption of light bulbs.[313] What followed was a campaign to accelerate the introduction of the compact fluorescent lamp, an alternative light bulb that uses 30–70% less energy.

In 1997, the U.S. Environmental Protection Agency (EPA) introduced the Energy Star voluntary labeling program.[314] This program established a benchmark for lighting performance and quality as well as bringing a clearly recognizable label to the marketplace. The goal was to offer consumers an efficient lighting option with no sacrifice in performance. In 2000, the UK's Energy Saving Trust introduced the Energy Saving Trust labeling and certification scheme for energy-efficient products in the UK.[315] Products that display this label have to meet strict criteria on energy savings.

The year 2006 was an important year of change. In the UK, a civil group called Ban the Bulb started a campaign to ban incandescent light bulbs. In the U.S., Walmart, the world's largest retailer, announced a marketing campaign to boost its sales of compact fluorescents to 100 million by the end of 2007, more than doubling its annual sales of such bulbs. That same year Philips, the world's largest lighting manufacturer, announced that it was going to stop marketing incandescent bulbs in Europe by 2016, and the European Lamp Companies Federation (the bulb manufacturers' trade association) supported a rise in EU lighting efficiency standards that would lead to a phase-out of incandescent bulbs.[316]

A year later, in 2007, Currys, the UK's largest electrical retail chain, announced that it would stop selling incandescent light bulbs.[317] That same year, the UK government announced that incandescent bulbs would be phased out by 2011.[318] And, in the U.S., the Energy Independence and Security Act (EISA) raised the minimum efficiency standards for traditional incandescent bulbs far beyond what the technology at that time could manage.[319]

Three years later, the market in the EU shifted to the fourth phase of market transformation. In 2009, the sale of 100W incandescent bulbs was banned in the EU. In successive years, sales of 75W and 60W bulbs were banned, and even the 40W is currently being phased out. The U.S. followed with similar policies only a few years later, in 2012.[320] Even China banned the 100W light bulb from the market in 2012.[321]

It is just a matter of time before traditional light bulbs will no longer be sold anywhere. This is simply because it is becoming less and less attractive for manufacturers to invest in this old technology. This market has shifted for good as more energy-efficient alternatives advance and energy-saving light bulbs have become the norm.

9.2 September 2025: a future scenario for agricultural commodity markets

We have seen how several other sectors and products have reached a level-playing-field situation. Now let's look at a potential fourth phase scenario in the year 2025 for the agricultural commodity sectors we have been following throughout this book.

> Ladies and gentlemen of the press, today we, the European Commission, have an important statement to make together with our partners from China, India, Indonesia, Vietnam, Brazil, and 15 other leading countries from Africa and Latin America, and leaders from the industries and civil society. Today, we have taken an important step toward the fulfillment of our sustainable trade relations and mutual ambitions. After four years of negotiations, a binding treaty has been signed to proclaim and recognize our mutual interdependency and need for collaborative action to provide sustainable food security for all. From now on we will take

> collective action on important issues that have threatened the future of our economies for too long. From this day forward our nations, industries, and value chains will work closely together. Consumers in the EU have the right to know where their products come from and to be assured that all of the food they consume is produced according to minimum safety and sustainability standards. The farmers and factory workers that produce these products are entitled to safe, decent, fair, and dignified working conditions. This, so we have vowed, is both a concern and a responsibility of us all...

It is a sunny afternoon in Brussels, more than ten years in the future. It is a day of celebration. After four years of intense negotiations behind the scenes, agreement has been reached between government officials, industry leaders, leaders from civil society (the NGOs), and research institutes from the EU and many key food-producing countries. The agreement states that, from now on, all food products imported into the EU will have to meet minimum criteria on sustainability, traceability, and food safety.

"It is the right and obvious thing to do. I don't understand why we haven't done this much earlier. Our consumers expect nothing less, they hold us accountable and we want to deliver on this promise," says Nestlé's CEO to media reporters. The Managing Director of WWF International and the Director of Greenpeace agree: "Indeed, it is about time. The industry and civil society have been jointly lobbying for this legislation for several years. It is fantastic to see that common sense has prevailed and has now become the law. Food is life, and it should be produced in a way that protects and enhances life."

The presidents of Brazil, Argentina, India, Indonesia, Vietnam, South Africa, and Nigeria have prepared a joint press statement:

> For far too long it has been common practice that the products produced in our countries were regarded as cheap and unsustainably produced. For too long we have accepted this situation and neglected the devastating effects this had on our people, our lands, and our ecosystems. As individual nations we have been unable to change these practices. Today, we unite together on a common cause. With this treaty stating clear objectives and with clear responsibilities, we know that we are all aligned and working toward the same objectives.

"Shake hands!" the reporters call out. On the stage a group of political leaders from different continents, industry leaders from multinationals, and leaders from civil society stand shoulder to shoulder and shake each other's hands. Cameras are capturing this historic moment. "Was it only 15 years ago that we were still protesting outside your doors and bashing your brands?" says the Director of Greenpeace to the CEO of Procter & Gamble with a smile. "Much has changed."

9.3 How it got to this point

It has not been an easy process. Trade and industry, civil society, and many other stakeholders have been lobbying for this treaty for a long time. The industry needed the EU to agree on sets of standards with the producing countries as well. This was important. Although the majority of the industry had been implementing and working in accordance to the new industry standards for a while, there were frequent reports of group of companies who deliberately continued to seek non-compliant, unsustainable farmers to gain cost advantages. With the EU policy and legislation in place, this behavior would no longer be allowed. The market would finally reach tipping point.

Another important reason why the European Commission needed to get on board was because of the national governments of the producing countries. For a long time, the industry negotiated with them, but they hesitated to set higher standards. Many of the national governments lacked the knowledge, the capacity and the resources to set up and run effective capacity-building and monitoring programs, to implement quality standards and provide alternative sources of income for the millions of farmers in their countries. Moreover, it took time before both the industry and the national governments trusted each other sufficiently to have this conversation and work together. For a long time, the big question was: Who will pay for all these investments in productivity, communications, education, and research? Was that a role for industry, for government, or for the multilaterals like the World Bank and FAO and the donor community?

Smart financing solutions eventually found the solution. As a way to pay for all the necessary investments in the agricultural sectors, and to build government programs, it was decided to impose a levy (fee) on every container not complying with the required norms that were exported from certain countries to EU ports. At the same time, products that did comply with the norms were granted VAT tax breaks so that sustainable products would not be more expensive in the supermarkets. In fact, the unsustainable products were starting to become more expensive. The levies were used to finance large support programs and government capacity-building programs for the more commercial farmers to increase productivity and to enforce new regulations. It proved to be a simple, yet effective, measure to reward the right behavior and punish the wrong behavior. Other than some protests and lobbying from individual companies that had never invested in sustainable practices and were now lagging behind, the great majority of the industry welcomed this policy.

Slowly, agriculture became a profession of choice again for young farmers in the emerging economies. With the new investments, better prices, higher standards, better support structures, and new technologies, the younger generation started to return to agriculture. Looking back, the sector has come a long way from when the market transformation process started.

9.4 Sustainability will become a mainstream qualifier

This clearly is a future scenario. However, if the market transformation curve is somewhat correct, and if it has any predictive value, then sustainability will become a mainstream qualifier and markets will have no other choice but to react to this reality. This is especially the case in agriculture.

It may sound unrealistic at this point this time, but a few years from now this scenario might start to look more and more real and obvious. After all, for many years the energy sector would never have believed that the incandescent light bulb would be banned from the market, and the mobile phone manufacturers would not have believed that they would have to provide a standard phone charger. Sectors and markets change, and what could be a better reason for innovation and market transformation than sustainability and the needs of future generations?

I predict that, in about 20 years from now, my children, who are currently below the age of ten, will not believe me when they read this book and when I explain to them how agricultural systems functioned in the first years of the new millennium. It will sound completely unbelievable to them that agriculture, the source of life, was once a sector caught in a global race to the bottom. Where hundreds of millions of farmers, who grew the food we all depended on, were living in poverty and where the production of food once equaled the destruction of complete ecosystems—the cradle of life. It will sound unbelievable because, in 20 years, people will care about and value where their food is coming from and how it is produced. They will care about the fate of the farmer who produced it. And because by then they understand, as Martin Luther King said in his "Christmas Sermon on Peace," that all life is interrelated. We are all caught in an inescapable network of mutuality, tied into a single garment of destiny. Whatever affects one directly affects all indirectly. And this is particularly true in agriculture.

9.5 Summary of the fourth phase of market transformation

Wouldn't it be great if the agricultural markets would follow a similar development path as we have seen in the egg, mobile phone, and light bulb sectors? When sustainability has become the norm rather than the exception? If governments support in the final transition phase, raise the bar and force players in the sector to comply or get out of the market? And if all this takes place in combination with a binding agreement between importing and exporting countries that would stop the race to the bottom, revitalize agricultural sectors, and provide sufficient, safe, and sustainable food?

The fourth phase is one of patience and perseverance. It does not have the energy and market dynamics of the previous phases, but it is a crucial phase nonetheless. It marks the finalization of a long change process in which the sector slowly grew to its current level of awareness, defined new rules of the game, and increased its level of connectability to deal with the problems and find alternative approaches. Finally, the new rules become part of the institutional and legal framework, and thus become mandatory for all.

But this isn't the end game. The change process doesn't stop here. Just as sustainability is not a goal to be accomplished but a journey to discover, new change is already on its way. As one definition of sustainability becomes normal for the entire industry, other issues, new insights, and new solutions and standards have emerged and they are already challenging this status quo. First movers have already rallied once again around new ideas to gain competitive advantage and, as they become successful, second and third movers will try to differentiate themselves from their competitors in the marketplace. A new transformation curve has already begun before the previous one has ended.

Summary of the fourth phase of market transformation: the level playing field

General characteristics	- Harmonized initiatives - Joint capacity building - Institutionalization - Involvement of national governments and international bodies
Level of awareness	- High level of awareness about the interconnectedness of the sector - How do we organize ourselves to change the rules of the game?
Perceived level of complexity	- This is the new normal: new topics emerge; the transformation curve starts over again on new issues
Main change agents	- Industry lobbies for a level playing field - Governments and trade organizations keep the rules - Law enforcement

Driving force for the market	- Compliance to mandatory standards is a qualifier for business
Who is against change?	- Laggard companies, national governments - Standards organizations
Dominant solution behavior	- Compliance with the law: see how we can compete in this new reality
Willingness to cooperate with others	- High: but as soon as the law becomes effective, competitive behavior increases again
How to get it going and how to be successful in this phase?	- Create an overarching vision for the industry - Get as much of the industry to align with, and contribute to, this new vision - Advocate, lobby, institutionalize
Limitations of solutions and barriers of this phase	- Although we have moved the sector, we realize that not all problems have disappeared, new issues arise, the curve starts again
How to transform to the next phase?	- The next phase is a new inception phase in which the new problems will be addressed

10 Key questions about market transformation

We have come to the end of the transformation curve. As we have seen, changing agricultural systems is an evolutionary process involving various stakeholders who eventually become more and more interconnected and increase their connectability to deal with the complexities that threaten their industry. Eventually, the rules of the game are changed and the "right behavior is rewarded" or even made mandatory while "the wrong behavior is punished." This is the complete opposite of the former situation that caused the problems in the first place.

Rewarding the right behavior: it sounds so logical, doesn't it? So why do we have to go through four painful phases and such a long change process to get there? Why can't we just skip a phase or two, jumping directly to the third or fourth phase, and change the system? And what happens if a phase in the transformation curve fails? Can the market transformation curve be applied to other complex problems in non-agricultural sectors as well?

These are good questions, which we hear a lot when we present the content of this book, and they are worth exploring, as they will expose some deeper layers of the transformation curve thinking as presented in this book. Let's address the 12 most frequently asked questions about the transformation curve and answer them in more detail.

The 12 most frequently asked questions are:

1. Phase 1 of the transition curve (Chapter 6) is about raising awareness and creating a sense of urgency, and most of the time NGOs have taken action to do this. Do we always need to raise awareness and create a sense of urgency to start a change process?
2. Can organizations other than NGOs also take on the role of awareness raiser?
3. Phase 2 of the transition curve (Chapter 7) is about standards and certification as the most effective instruments for change. Does it always have to be standards and/or certification?
4. Are (sustainability) standards no longer necessary after phase 3?
5. Is there only one transformation curve or are several curves going on at the same time?
6. What happens in phase 4? Have we reached true sustainability, and does the change process then stop?
7. Is it possible to skip phases or speed up the change process?
8. Can a transformation phase fail? If so, why can it fail and what happens then?
9. Does a sector move as a whole from one phase to the next, or are there groups of leaders, followers, and laggards, and does this differ per region or market?
10. What is the role of government in this transformation curve?
11. Can this transformation curve also be used to solve complex problems in other sectors?
12. What can individuals or organizations do to initiate change to solve a complex issue they care about?

Let's go into more detail on these questions.

10.1 Do we always need a certain level of urgency to start the change process?

The answer to this question is "yes," there must always be a high sense of urgency or a "burning platform" for people or companies to change. Particularly in the case of systemic change, which means that multiple organizations need to change their behavior in relation to each other and in relation to a higher objective/ambition, in order to solve the problem. This sense of urgency is needed, not only because it is human nature to resist change; people in general tend to choose the familiar and predictable over the new. It is also because, from the viewpoint of each individual actor within the system, there are perfectly good economic and logical reasons to continue doing business as usual for as long as possible. After all, most organizations within the "old" system have invested a lot of time and resources in becoming successful according to the rules of that system and it is working for them. So why change it?

Consider the example of the rabbit population from Chapter 3, which can over the long run only remain healthy and vital when the reinforcing and balancing loops are in balance. Imagine what happens when no more predators are hunting them. That is good news for the rabbits, you might say. The population expands rapidly. During that period of rapid expansion, the rabbit population seems to be well functioning and highly successful; after all, they are reproducing and the rabbit community is growing fast. But at some point the vegetation can no longer sustain the rabbit population, or diseases break out. At some point in time, this unfortunate population inevitably hits the limits of its supporting system, and crisis strikes.

Our markets can, in many ways, be compared to a rabbit population without balancing loops. Growth is for us the number one success indicator, and we too hit the limits of our support systems. Examples of these systemic limits in our markets include massive fires in tropical rainforests, poverty, banking crises, climate change, or resource scarcity. However, for a sector in phase 1 of the transformation curve, these signs are often not seen as signals of market failure, but rather as indications of an efficient and well-functioning, successful market: "This is how business is done"; "We are growing"; and, "Why change a winning formula?" From a systemic point of view, humans are apparently not that different from rabbits.

But these crises that happen do raise awareness that something needs to change and they force players to move; that is exactly what is needed to start the transformation curve. For any significant change to happen, a high sense of urgency is needed to wake up the system and prepare the actors for change.

10.2 Are NGOs always the first to sound the alarm and raise awareness?

No, it is not always NGOs who sound the alarm or raise awareness. In fact, anybody can start this process. In some cases companies play this role; they recognize that their industry is running into a wall and see an opportunity if they can change before their competitors. In other cases, the awareness process is started by governments, multilaterals, concerned individuals, entrepreneurs, or scientists and researchers. There are hundreds of cases where systemic change happened and was initiated by individuals or organizations who cared about the issue.

To raise awareness, you must first and foremost care deeply about the issue at stake, have some credibility, and not be dependent on, or feeding off, the unsustainable practices in the first place. This is why, in many cases, the change in the first phase of market transformation is initiated by people or organizations who are relative outsiders to the system; they are not dependent on it. Organizations such as NGOs are generally not part of, or dependent on, the current industry. They are considered by many people as credible, their constituency cares deeply about the issue, and they are self-mandated to act against the calamities that are occurring. That is why, in many cases, NGOs play this important role of awareness raising.

10.3 Does it always have to be standards as instruments of change in phase 2?

No, other instruments can have a similar function. But the new instruments have to become an accepted standardized way of doing things in a new way in order to have the systemic effects. Sustainability standards and certification programs as described in this book are successful as instrument of change for two reasons. First, they establish an accepted, standardized definition between various organizations on what sustainability is, or should be. Second, they reward organizations for good behavior (i.e., for value chains to change their sourcing practices and farmers to change their farming practices). There is value in doing the right thing, and this enables first movers to become successful. This, in turn, triggers competition between the first, second, and third movers, as they will imitate, but never follow. This is a very powerful dynamic.

Other best practices, concepts, or methodologies can also start the competitive race to the top. These include microfinance (which found a standardized way to create value for both the lender and the recipient of the money), the Dow Jones Sustainability Index (which is an accepted standardized way of ranking companies on their sustainability behavior and creates benchmarks), or new accepted and standardized technologies to help consumers verify the authenticity of the medicines they buy. Another example is an accepted, standardized rating mechanism to assess the bankability of farmers and link them to first-mover banks and input providers.

All of these practices and concepts have one factor in common: they standardize their approach in an accepted way and create value for first movers to change their behavior. This sparks competition so that others will change theirs as well. In agriculture, sustainability standards and certifications programs have fulfilled this important function effectively.

10.4 Do standards play a role in and beyond phase 3?

This is an important question that is asked a lot these days. The answer is that it depends on the situation. Sustainability standards have proven to be an effective instrument to start the second phase of market transformation. At some point, however, competition on standards no longer serves nor solves the issue, as we have seen, so the industry has to move beyond standards as the main instrument of change. Companies and governments have to start working together, and address the non-competitive causes of market failure (as described in Chapter 4, the three loops of market failure are: failing markets; failing governments; and the absence of conditions for change). Currently, the standards and certification business case is the non-trust issue between and amongst market players, governments, consumers, and other stakeholders. As the industry matures throughout the transformation curve, the level of trust and connectability slowly increases, diminishing the business case for external standards and certification programs. Finally, the new norms will need institutionalization and legislation. That is when standards have served their ultimate purpose as instruments of change and they either have to move upwards, differentiate, merge, become a service provider, or leave the marketplace.

Besides the need to move beyond standards as the main instrument for change, standards start to lose their brand differentiation value in the market as well. Currently, there are over 500 sustainability labels on the market, which causes confusion, lack of transparency, and ultimately a breakdown of trust and credibility with consumers, farmers, governments, and companies. Each additional label entering the market further dilutes the differentiation factor of the other labels. This causes the value (marketing, media, or brand value) of all labels to diminish, while the recurring costs (e.g., certification costs) of these labels and standards remain high. Finally, we have seen that national governments in some cases reject Western standards, which they view as a breach of their national sovereignty.

While the market progresses from one phase to the next, and the focus in the industry shifts from competition to collaboration, standards organizations (certifiers) have to reflect on the following questions: "What is our function in the marketplace and what is our added value?" "Does the industry always need labels on a package; does it always need that third-party credibility?" "Is it possible that one day the industry will define its own standards and implement them, and that compliance will be checked by their other verifiers or auditors or even the government, as it should be?" I suggest that these are the right questions to ask and the answers to these questions will determine the future of standards and labels.

Many of the current standards and labels will disappear over time, while other, more efficient and effective sets of indicators, practices, and verification methodologies will emerge and take over. I believe this is a natural part of evolution and progress. This is how it should be.

Just as we saw with some NGOs, who go back to earlier phases of the transformation curve and pick up their function of campaigning and awareness-raising on new issues, perhaps the standards organizations could find new issues that need standardizing and first-mover competition. After all, that's their natural role in the larger ecosystem.

10.5 Is there one transformation curve or are there several going on at the same time?

Part II of the book describes the changes that occur in sectors such as palm oil, cocoa, and coffee as completely distinct and clean phases of a single transformation curve. In reality, though, more than one curve is always going on at the same time. In every sector, at any given moment, various change processes are ongoing, each in a different phase of its own transformation curve. Like waves on a beach, these transformation curves continue to come, one after the other.

At any given moment there are new, challenging, ideas with the potential to be successful. During phase 1, these new ideas compete with each other for survival. In phase 2, the new emerging best practices compete with each other for first-mover advantages and struggle against the status quo in the industry. And in phases 3 and 4, the winning best practices are institutionalized in a collaborative effort between industry and government. By definition, each phase is constantly being challenged by new concepts and practices that are in earlier phases of development.

We can observe this process in other sectors, such as the market for solar energy. Numerous countries are currently in phase 3 of market transformation, where governments and industry are working together to build an enabling environment to lower the price of solar panels, and stimulate the large-scale uptake of solar panels on roofs. This is the result of previous phases of market transformation during which solar energy competed for many years with other alternative-energy technologies and battled against the status quo of the fossil fuel industry.

While the enabling environment for solar energy is currently being built (phase 3), at the same time solar energy panels themselves are being challenged by a constant stream of new ideas claiming better performance, or even suggesting that solar energy is an old technology and that the industry should embrace a different, disruptive, set of best practices, such as hydrogen energy from waves or biodiesel from algae. The latter industries are currently in phase 1 or 2. Hence, while a sector is getting organized, institutionalized, and settled around a single best practice that has risen above the ranks in phase 4, it will, at the same time, be challenged by other competing ideas that seek to replace it.

10.6 Do we reach true sustainability in phase 4, and does the change process then stop?

Phase 4 of market transformation represents a big step for the whole industry, but that does not necessarily equal real sustainability, and it certainly does not represent a steady end state. It is merely a certain period during which industry and government have institutionalized and codified the consensus on best practice. It is, at that moment, the new normal. As stated in the answer to the previous question, the waves of change continue to come.

First, when a certain best practice has reached phase 4 and has become a "qualifier," industry can no longer compete and differentiate on this particular aspect. Naturally, some first movers will seek newer, better, or more demanding practices to gain a competitive advantage. This is the same competitive force that got them to become a first mover in the former transformation curve earlier. An example of this is when barn eggs became the norm in the EU. Some large industry players, such as Unilever, who had used the claim they used barn eggs in their products, quickly scaled up their sourcing ambitions from barn eggs to free-range eggs, which is a slightly higher standard on the animal welfare scale for chickens and allowed them to continue to use this differentiating claim on their products.[322]

Second, although industry has just accepted a new norm that moved it upwards and leveled the playing field for all players, a new crisis or major event is bound to hit the sector, and the cycle will start all over again. NGOs or other change-makers then use the crisis to raise awareness, and a new first phase of market transformation begins.

Systems will continue to change. And they will continue to move through the different phases of market transformation.

10.7 Can you skip a phase or can accelerate the systemic change process?

This book presents a four-phased model. But in some cases the second phase is skipped, or not relevant. In cases where all three loops of system failure are present—failing markets, failing government, and where the conditions for change are absent—the chances are that the sector will go through all four phases, with each phase a necessary preparation for the next one. Under these circumstances I believe you cannot really skip a phase.

This is different to situations where governments have a much more direct influence on these issues, alternatives are available, and they can simply forbid a practice. The example of the harmonization of mobile phone charge in the EU market falls into this category. In that case, the second phase of market transformation, that of first-mover competition on alternatives, is skipped.

In both cases, however, you cannot grow from a baby to an adult overnight. Systemic change happens through a process of maturation.

More important is the question of whether you can accelerate the systemic change process. Yes, I do believe this is possible and also very necessary. By taking the right measures at the right time and by actively organizing and structuring the landscape around the complex issue that needs solving, it is possible to evoke, accelerate, and facilitate the change process. The flipside is also true. By taking the wrong measures, by stimulating the old thinking and interests, by not having a vision of systemic change, or by doing nothing, it is possible to slow down a systemic change process, or even cause it to fail. There are many cases where this has happened.

In each phase there are specific interventions you can take to speed up the change process:

- Phase 1: help to initiate and raise the level of awareness and the sense of urgency in the sector
- Phase 2: recognize new best practices that enter the market, help them to get standardized, reward first movers, and intensify the level of competition between first, second, and third movers
- Phase 3: facilitate and incentivize the willingness to go beyond competition and to move toward collaboration. This requires neutral convening, facilitating dialogues with governments, smart and shared financing mechanisms, and creating an enabling policy environment
- Phase 4: help to institutionalize the change and create a level playing field

Just as each phase has its own dynamics and characteristics, so does each phase require different interventions and tactics, and these interventions can be actively organized and managed. They are preferably introduced and facilitated by neutral, credible, change agents that help the change process move more effectively.

An interesting observation is that once one sector has paved the way from one phase to the next, this knowledge seems to spread around to other sectors quickly and they will follow the leader. So if one sector moves it seems to accelerate the change process of other sectors as they will follow. We have seen this with the coffee sector. It was once a pioneering sector, and the birthing place of all kinds of innovation and standards that started in competition with each other and brought the sector to the second phase of market transformation. In turn, the coffee sector had learned a lot from the innovations that had taken place earlier with sustainability standards in the forestry sector and the fishing industry.

Not long after these coffee standards became successful, many other sectors followed the coffee example and the same dynamics of competing standards was introduced as well. This explains why almost all sectors are currently in the second phase of the transition curve. In coming years we will most probably see some of the sectors following the coffee and cocoa example and moving to the third phase as well.

A specific remark about the role of government: the law-making and policy-making function of government becomes particularly important in phases 3 and 4. However, governments can make a big difference in phases 1 and 2 as well. These phases, as we have seen, are all about developing and starting competition on new and disruptive ideas. Unfortunately, for many governments, working with new and disruptive ideas is way out of their comfort zone. They would rather continue to deal with the safe and the known than with the new and risky.

When they first appear the new ideas are, by definition, new, untested, risky, they challenge current thinking, and they go against the interests of the current system. As a natural result, the initial acceptance of new ideas is low and resistance against them is high. It is a fact that the status quo of the old system will resist any new ideas that may disrupt their model. Most of them would like to see these new ideas fail so things can go back and stay "normal." We see this everywhere. Telecom providers, for example, fought and are still fighting tooth and nail against free text messaging or free phone calls over the internet. The fossil fuel energy companies are lobbying against alternative energy sources. Banks are against peer-to-peer lending mechanisms. Software companies are trying to stop open source software. This is normal. We should be very careful to judge new ideas on their initial acceptance by the old system. That would be a classic and very expensive mistake.

It is in this early phase that support and an enabling environment for these new ideas are important. New ideas need an environment where they can prove themselves and improve, where they can find resources, support, and recognition and access to networks. They need a different approach and environment than an already tested and proven concept. However, not many governments and civil servants in these governments know how to deal with new, untested, risky ideas—precisely because they are new, untested, and risky. It is much safer to continue giving support to the "old" system of industries, universities, research institutes, multilaterals, etc., which hampers innovation.

If governments are serious about being a partner in solving complex problems and market transformation processes, then I strongly advise them to think about how to co-create with others an environment where new, disruptive ideas can take shape and grow, and how they can give recognition and support to innovations. We must stop trying to solve the problems of tomorrow with the solutions, institutions and business models of last week. Instead, we can incentivize and stimulate social innovation and entrepreneurs[323] to find new, disruptive ways. Indeed, many of these new ideas will fail. That is inherent to innovation. But by providing a more supportive environment, and a recognition that just because it is new it doesn't mean it won't work, we can increase the success rate and decrease the mortality rate of good ideas and frustrated entrepreneurs; we need only a few successful ideas to trigger the second phase of the transformation curve. These new ideas will ultimately change the world much more efficiently than continuing to do what we have been doing for the last ten or twenty years.

10.8 Can a phase fail? If so, why, and what happens then?

Yes, each phase of market transformation can fail or slow down—temporarily or in some cases almost more permanently. And many of them do fail or slow down. Reasons for failure differ per phase.

- *Reasons for failure in phase 1*. The sense of urgency is not high enough and therefore the willingness to change is too low. Or the push back from the old system is too big and any attempt to introduce new thinking is stopped. In that case "business as usual" takes over again. A notorious example is the financial sector. After the financial crisis of 2008 and the bank bailout, governments and private citizens reacted in horror. Politicians demanded harsh penalties, and private citizens protested against greed and inequality through the international Occupy

Movement. Now, five years later, it is safe to say that not much has changed in the financial sector. It remains stuck in phase 1 of doing symbolic projects rather than taking it to the core of their value proposition. I believe the change process in the financial sector has failed and a new crisis awaits.

- *Reasons for failure in phase 2.* The new practices that are emerging do not create sufficient value for first movers to get them to act and compete. Instead, companies continue with their own projects and stay locked in this first phase. Another reason for failure could be that organizations which benefit from the project phase successfully resist, undermine, or lobby against the emergence of new practices. A lack of resources, or unwillingness to pay, is also a known fail factor in this phase.
- *Reasons for failure in phase 3.* Companies start to work together, but do not open up and grow with the collaborative and noncompetitive movement. Or companies, NGOs, and standards organizations can successfully resist and undermine collaboration between actors. Another reason could be: if companies don't trust each other sufficiently, then they won't really collaborate and instead will stay in the competition phase. This phase can also fail if governments do not provide leadership and demonstrate their willingness to change the policy landscape to institutionalize better practices. A lack of resources or unwillingness to pay to make the change mainstream is another reason for failure.
- *Reasons for failure in phase 4.* The final phase can fail if governments are unwilling to make institutional change, or if there is effective lobbying against change. This phase requires political leadership and that is not always present.

Although the market transformation processes may fail or get stuck in a certain phase, this doesn't mean that change will not happen. The root causes of the failing system are still there and it is only a matter of time before problems return, probably worse than before. Eventually, the case for change will become more urgent and more pressing than ever, and the sector will have to change or become obsolete.

10.9 Does a sector move as a whole from one phase to another, or are there groups of leaders and followers? Does this differ per region or market?

A sector never moves as a whole. There are always leaders, followers, and laggards. For example, the coffee sector is in phase 3 in Europe, but it is still somewhere between phases 1 and 2 in the U.S. The same is true in producing countries, where in some regions the acceptance of standards and sustainable production methods is much higher than in others. For example, we see a large uptake of responsible palm oil in Europe, but in South-East Asia the demand for sustainable palm oil is negligible. This difference in uptake of the change process eventually becomes a major obstacle to achieving a level playing field (phase 4). For example, it is very hard for all companies in the EU to come to an agreement about buying only sustainable products if their competitors in Asia have no qualms about buying cheap and unsustainable products. This may lead to a situation where the wrong behavior is rewarded once again.

Market transformation processes are always driven by a nucleus of leading companies and stakeholders. This provides the force that moves a sector forward. At a later stage, if the change process is to become successful, other companies or countries must follow these leaders into the next phase.

10.10 Is this transformation curve also useful for solving complex problems outside of agriculture?

Yes, I believe it is.

Since the root causes that create problems in agriculture on a global scale are similar to the causes of many other complex problems, I believe that the solution will be similar as well. Climate change, food scandals, financial crises, corruption, environmental problems, plastic waste in the oceans, over-exploitation of resources, and almost all other complex problems are driven by the same rules of system failure. In Chapter 3 these were identified as:

1. Actors in the system are driven by short-term self-interest.
2. There is a large distance between actions and consequences. Actors can act anonymously and there is no mutual interest in each other's wellbeing. Individual actors can shift negative consequences to others in place or time. In other words, the costs can be externalized, and those who do this are the winners.
3. No authority exists that is present, willing, or able to correct or counteract the negative dynamics.
4. Any serious solutions are stuck in complex chicken-and-egg situations, because the conditions for change are not present.

If we look around us we see situations where these four elements are dominant; the result is a culture where the wrong behavior is rewarded and the right behavior is punished. By our collective actions, the wrong incentives have created complex and widespread global problems. The solution can only be found in creating a higher level of interconnectedness and connectability, and by changing the rules of the game.

Therefore, when there is a market situation that acts on self-interest in the short term, with the absence of a strong authority that can impose rules and standards, I believe the transformation curve as presented in this book is an effective model. This approach can help to determine the current phase of a market in relation to the problem, and can help identify which interventions are necessary to bring the market closer to a structural solution.

10.11 Who are the leaders of the transformation? Do we have the same leaders throughout the change process?

Each phase has its own leaders and heroes, even though most of these people would not call themselves leaders or heroes. They would say they are simply doing their job or are following their heart. In the first phase the leaders and heroes are the campaigners and awareness raisers. They could be NGOs, research centers, concerned citizens, students, critical media, entrepreneurs, innovators, or many others. They stand up against common beliefs and accepted norms, and face resistance and ridiculous counter-arguments. Then there are the organizations that are willing to accept that there is a problem and start projects to experiment and find solutions. They are also heroes of this phase, albeit less glamorous. In hindsight it may be clear that their initial actions did not solve the problem, but together with campaigners they are creating the preconditions for phase 2 of the transformation curve.

In phase 2, the heroes are the new practices, such as standards, and their first movers. In particular, the first movers shake the industry. They boldly go where none of their peers has gone before. Often driven by enlightened self-interest, these first-mover companies embrace change and create value from it. This first-mover value triggers competition which creates unparalleled market dynamics.

The leaders in phase 3 of market transformation are the companies who dare to question the competitive model that brought them to where they are, and who now initiate non-competitive collaboration. It takes courage and leadership to accept that the problems you face as a multi-billion-dollar company are too big for you to handle alone, and that you need to work together with your competitors to develop and share knowledge. The change agents behind the scenes, the neutral platforms and facilitators, who are driving the change and who are absorbing and dealing with resistance, are also vital for the transformation.

Moving toward phase 4, the leaders are those national governments who are willing to become part of the solution, change their own policies to effectively implement new practices, and make these practices the norm. Sometimes these leaders encounter fierce lobbying and political pressure.

Throughout the transformation curve, the agendas, drivers, and interests change. The leadership changes as well. In all cases, people, companies, or organizations that show leadership deserve our recognition and support. By recognizing their leadership and contribution we can speed up the systemic change process and improve the success rates.

10.12 What can I do to initiate change on an issue I care about?

Everyone is a change-maker.[324] I truly believe that this to be true.

Whether you work for a large multinational, a small organization, government, an NGO, a research institute, the media, or if you are simply a concerned citizen, there is much you can do to be part of the transformational change.

Each phase of the transformation curve calls for leadership—systemic leadership to look beyond the short-term interests for yourself or your organization and to stop externalizing the costs to others. These leaders ask: what is the right thing that I can do now for the system at large? In a world where the wrong behavior is rewarded, most people will do what is easy, instead of doing what is right. Therefore, this kind of systemic leadership is rare in both the corporate and political arenas, as it feels as though you are going against your own interests. Not only that, you may also have to face severe resistance, criticism, and cynicism, and even in some cases violence. Nevertheless, real systemic leaders decide to stand up against all odds and face the failing system. This is how change happens, every time.

These change-makers need support; luckily, a number of platforms do support them. To name a few examples: Ashoka Fellows Network; The Forum of Young Global Leaders, the Global Shapers, and the Schwab Foundation for Social Entrepreneurship of the World Economic Forum; Echoing Green; the Unreasonable Institute; and UnLtd.[325,326,327,328]

In many cases, it is the combination of the heart, courage, and entrepreneurship of the change-maker and the supporting platform that is able to make it through all the levels of resistance and become successful. Without these individuals and support platforms, there would be less change, and we would face far more complex problems in the future.

As a tribute to these change-makers we will present ten inspiring examples of individuals or small organizations that are working on systemic change. Each of them is a true systemic leader who is changing the world, but nobody will know them for the great work they are doing. For most of them are not looking for fame and glory; they just want the issues they care deeply about to change and improve.

In total there are thousands more of these change-makers. Within their own capacity and within their communities or regions, one day they decided that they wanted to no longer accept the situation and to change the rules of the game. Each of them is going through similar phases of the change process and each of them is using similar tactics as they go through the market transformation process. Every day they prove that systemic change is possible. It may not be easy, nor can it be done quickly, but it can be done.

Not all of them have university or business school degrees. In fact, most of them do not. They do not have many resources, except for their imagination, their courage, and their determination. And they did not have much support in the beginning. In fact, they faced resistance, indifference, and cynicism—and sometimes threats or violence toward themselves or their loved ones. Despite all these challenges, they still decided to do the right thing.

It is my hope that, by reading these examples, more people will feel inspired to take on issues they care deeply about and will change the system that is at the root of those problems. And it is my hope that more platforms, organizations, and governments will help these people.

I end this book with the same quote I started with because, for me, it captures the heart of systemic leadership and sustainable market transformation.

> "Never doubt that a small group of thoughtful, committed citizens can change the world; indeed, it's the only thing that ever has."
>
> Margaret Mead, U.S. anthropologist and popularizer of anthropology (1901–1978)

11
Ten examples of inspirational change-makers

Topic	Name (organization)	Location	Period
Aerial spraying of agrochemicals	Sofía Gatica (Mothers of Ituzaingó)	Argentina	1990s to present

Link to tough problem

During the economic crisis in Argentina in the 1990s, the national government began to promote genetically modified soy. These new soybeans were developed by Monsanto in the mid-1990s. They were engineered to resist herbicides, which helped farmers to improve their productivity. Farmers welcomed the introduction of genetically modified soy, as it helped them to remain competitive in a sector where economies of scale are important to reduce production costs per acre.[329] Now, 20 years later, the production of soy in Argentina has taken a big leap. For example, taxes on soy exports are responsible for 6–9% of Argentina's total tax receipts. The use of herbicides, however, has dramatically surged.

Action to make a change

Meanwhile, in 1999 in a suburb of Córdoba surrounded by soy fields, Sofía Gatica's fourth child died three days after she was born due to kidney failure. After the death of her daughter, Gatica noticed that more children were being born with deformities and that many people were developing cancer. So she began investigating what was going on in her community, and started a door-to-door survey about health problems in her neighborhood. This was the start of the project and awareness phase.[330]

Current market transformation phase: beginning of phase 3

Sofia was soon joined by other concerned mothers, who called themselves the Mothers of Ituzaingó. The results of the survey showed that cancer rates were 41 times the national average, and the Mothers of Ituzaingó demanded an immediate investigation. In 2002, the government agreed to investigate, which led to shocking conclusions about water pollution and the presence of herbicides in neighborhood children's blood. Although they were physically and verbally threatened by farmers and by individuals believed to be paid by Monsanto, the Mothers of Ituzaingó continued their campaign, and support increased slowly. The pressure for action also increased when similar cases were reported across Argentina. In 2009, a doctor at the University of Buenos Aires (Andrés Carrasco) found a relationship between the birth deformities and certain components in the herbicides. In 2012, Gatica managed to get a municipal ordinance enforced which prohibits aerial spraying of herbicides within 2,500 meters of people's homes. Furthermore, in July 2013, the insecticide endosulfan (a very toxic insecticide) was banned in Argentina and many other countries.[331]

Future scenario

Despite the economic importance and the world's growing demand for food, Gatica and the Mothers of Ituzaingó continue their work to further control the activities of soy farmers in terms of herbicide and pesticide use. She wants genetically modified soy and all associated chemicals banned. At the same time, many farmers remain convinced that the continued use of Monsanto's products is crucial for Argentina's economic stability and its future food security.

***

Topic	Name (organization)	Location	Period
Corruption	Anna Hazare (India Against Corruption)	India	2000s to present

Link to tough problem

Corruption is everywhere in Indian daily life. For instance, government certificates for every life event (marriage, death, and many others) require, in many cases, a bribe, and the police are notorious for a wide range of questionable practices.[332] Several factors have contributed to India's high corruption score, including high tax rates, poorly paid government officials, and excessive regulation. As a result, corruption negatively affects the entire economy; the World Bank has identified corruption as the single greatest obstacle to economic and social development.[333,334] This case is a clear example of a systemic problem, as individuals are driven by short-term self-interest, authority is failing, and there is a large distance between actions and consequences.

Action to make a change

India has a legal framework in place that should reduce corruption. Anti-corruption laws include the Prevention of Corruption Act (1988), Prevention of Money Laundering Act (2005), and Right to Information Act (2005). However, there is a huge gap between anti-corruption policies and practice.[335] Therefore, numerous initiatives have been launched which aim to limit corruption in various ways. A campaign that attracted worldwide attention was the India Against Corruption (IAC) campaign. The IAC was formed in 2006 by Anna Hazare, a social activist, popularly known as "Team Anna." The campaign was launched in 2011, requesting stronger legislation and law enforcement against corruption.[336]

Current market transformation phase: moving to phase 4

The IAC proposed to install an independent body that could monitor and arrest corrupt politicians (the Jan Lokpal Bill). Another aim was the repatriation of "black" money from Swiss and other foreign banks. The IAC campaign gained considerable momentum and mobilized millions of people when Anna Hazare, started a hunger strike in April 2011. Protests spread throughout the country and eventually civil society organizations, together with the Indian government, started to work on a draft version of the Lokpal Bill. Meanwhile, competition among civil society groups started on who should lead the campaign, resulting in fragmentation.[337] Finally, after several amendments, the Lokpal and Lokayuktas Act (2013) came into force on January 16, 2014.

Future scenario

It seems unlikely that India's corruption level will decrease significantly in the near future even with the Lokpal and Lokayuktas Act in place, although it is a huge step forward. Critics have noted that corruption is deeply rooted in Indian society, and that civil society has reduced the entire issue to a law and order problem.[338] This is also shown by a KPMG Bribery and Corruption Survey held in 2011 among corporate employees. When asked what the scenario would be for the (by then) next two years, 46% of respondents said that the level of corruption would remain at the same level, 15% expected an increase irrespective of the legislation, and only 6% expected a drop of more than 25%.[339] Nevertheless, small steps are being made because of the efforts of individuals that force the government to implement and enforce anti-bribery laws.

***

Topic	Name (organization)	Location	Period
Waste management	Bambang Suwerda (Public Health Workshop)	Indonesia	2008 to present

Link to tough problem

Changing consumption patterns, uncontrolled population growth, and rapid urbanization have led to excessive generation of urban solid waste in Indonesia.[340] It has been estimated that the total amount of solid waste produced in 2010 was approximately 65.9 million tones, which is double the amount produced in 2006.[341] This affects Indonesian city residents in a number of ways, including deterioration of their physical environment, degradation of health and sanitation, and long-term alteration of environmental mind-sets.[342]

Action to make a change

A number of (unsuccessful) initiatives have been started to address this problem, both by local governments and environmental groups. As a result awareness has increased, but the problem itself has not been solved. Therefore, Ashoka Fellow, Bambang Suwerda, together with communities, established "waste banks" that function as community-based recovery facilities. The waste banks collect materials from clients, resell them to buyers, and add the revenues to the clients' "waste banks savings account." The program is accompanied by Suwerda's efforts to increase youth participation by integrating waste management practices in school curricula.[343] By giving the people money for their waste the system dynamics change, as it creates an incentive for people to collect and recycle their waste.

Current market transformation phase: phase 3

Suwerda's project has been very successful. The waste that is dumped into garbage disposal sites has been reduced, additional jobs have been created, and illegal dumping is eliminated. In total, the introduction of the waste banks has resulted in a 30–40% reduction of plastics waste generated across the district. The Ministry of Environment in Indonesia has recognized the program's success and has adopted and replicated the model nationwide. Another effect of the waste banks is a change in the public perception of waste management, which has fostered other local recycling businesses.[344]

Future scenario

To refine the waste bank model, Suwerda established a research and development center, called the Public Health Workshop. Furthermore, as the model is being used more and more throughout the country, Suwerda has focused on standardizing the requirements and procedures for operating waste banks. Eventually, Suwerda wants every village to have an operative waste bank.[345]

***

Topic	Name (organization)	Location	Period
Soil erosion	Hayrettin Karaca (TEMA Foundation)	Turkey	1992 to present

Link to tough problem

The Turkish Foundation for Combating Soil Erosion, Reforestation and the Protection of the Natural Habitat (TEMA Foundation) stated that, as a result of climate change and people's behavior, 743 million tons of soil is lost annually due to erosion in Turkey.[346] This means that 90% of Turkey's land is affected by soil erosion, and 50 billion m^3 of water is lost annually, as it can no longer be stored in the soil.[347] As a result, agricultural productivity is decreasing, forcing hundreds of thousands of rural people to migrate to the cities, with disastrous consequences. Although the erosion can be partly attributed to climate change, humans also cause and catalyze the process due to antiquated farming techniques, industrial and urban expansion, and destruction of forests.

Action to make a change

Hayrettin Karaca managed to put the issue of soil erosion on the Turkish public agenda in the 1990s, including the role of the human factor in this problem. In 1992, together with a friend, he started the TEMA Foundation. To address the issue of soil erosion, Karaca introduced a number of environmentally friendly alternative-income opportunities to reduce agriculture's impact on soil erosion, such as good agricultural practices, eco-tourism, and beekeeping. Meanwhile, he combined these livelihood projects with advocacy and education programs to raise awareness among the Turkish people.[348]

Current market transformation phase: phase 3

Karaca has made soil erosion, and the environment in general, an issue of public concern. In terms of legislation, TEMA was the driving force behind Turkey's ratification of the UN Convention on Desertification in 1994, the ratification of the Pastures Act by the Turkish parliament in 1998, and the enforcement of the Soil Conservation and Land Management Act in 2005. Moreover, since its start in 1992, TEMA's educational programs have reached more than 2.5 million people; today, 64% of adults consider erosion to be an urgent threat in Turkey. Finally, TEMA currently has 450,000 volunteer members, clubs in 60 universities, 550 voluntary representatives and over 100 scientists. It has also expanded to Germany, Belgium, and the Netherlands.

Future scenario

Over time, as land becomes scarcer, more and more people will become aware of the situation. Eventually, this will result in more, and more stringent, legislation for farmers. Step-by-step, a level playing field will be created. Although several crises, landslides, and famines will probably be needed to actually create this level playing field.

***

Topic	Name (organization)	Location	Period
Drunk driving	Candy Lightner (Mothers Against Drunk Driving)	U.S.	1980 to present

Link to tough problem

For decades, drunk driving was not considered a major issue. It was not something that people were proud of, but it was something that everyone did, and when it happened people tended to look in the other direction. Much has improved since those days. In 1982 there were 26,173 alcohol-related auto fatalities in the U.S., compared to 12,774 in 2009.[349]

Action to make a change

In 1980, Candy Lightner founded Mothers Against Drunk Driving (MADD). She started this initiative the day after her daughter's funeral. A driver with too much alcohol in his blood had killed her child. Lightner wanted to raise awareness on the dangers of drunk driving and fought for stricter laws against intoxicated driving. In order to raise awareness and to get people to care, she used an effective tactic. She showed photos of the victims' faces to make people realize that real people are behind cold statistics. What if this happened to somebody you love? Through continued campaigns, more and more people became persuaded that alcohol and driving are an unacceptable combination.[350]

Current market transformation phase: phase 4

After raising enough awareness on a small scale, MADD was gaining momentum. MADD's campaign was featured in high-profile TV programs in the U.S. Lightner spoke to many groups throughout society and even addressed Congress. The intense campaigning worked and MADD managed to get legislation implemented.[351] About 700 new drunk driving laws have come into force between 1980 and 1985, and MADD was a very important factor in this process.[352] Moreover, drunk driving has become more and more socially unacceptable. Although there are still people in the U.S. who think they have the right to drive drunk, it is much more demonized now than it used to be.[353]

Future scenario

The case of drunk driving is clearly in the fourth phase in the majority of countries around the world. Laws have been implemented and it has become socially unacceptable among a large part of the world population. Nevertheless, every year drunk driving is still responsible for many fatal accidents. In order to solve the problem the root causes of excessive drinking need to be addressed.

***

Topic	Name (organization)	Location	Period
Plastic waste	Charles Moore (Algalita Marine Research Institute)	Multiple countries around the world	1997 to present

Link to tough problem

Charles Moore, a researcher and sailor, is fighting plastic pollution. His journey began in 1997 when Charles and his sailing crew returned from the Los Angeles to Hawaii Transpac sailing race. They didn't know it yet, but they were about to make a grand discovery. While sailing through one of the most remote regions of the ocean, they saw plastic debris floating. Enormous amounts of it. They discovered the North Pacific Gyre Garbage Patch, an area the size of France, Spain, and Italy combined, filled with plastic waste.

"As I gazed from the deck at the surface of what ought to have been a pristine ocean," Moore later wrote in an essay for *Natural History*, "I was confronted, as far as the eye could see, with the sight of plastic. It seemed unbelievable, but I never found a clear spot. In the week it took to cross the subtropical high, no matter what time of day I looked, plastic debris was floating everywhere: bottles, bottle caps, wrappers, fragments."[354]

The plastic waste collects on this remote site because the circular ocean currents bring it to this point, rather like a huge waste collection point. And since plastic is not biodegradable it continues to accumulate. Something needed to be done. And Moore took action.

Action to make a change

He radically changed the focus of the Algalita Marine Research Institute (AMRI), which he had founded in 1994.[355] Since 1997, the institute has focused on a better understanding of the magnitude of our plastics "footprint," including the effects of fish ingestion of plastic on human health. Besides conducting research, Moore started to raise awareness. In a 2009 TED talk he explained the problem to a worldwide audience. Other parties started to take up the problem of plastic waste. In 2009, the UN released the report *Marine Litter: A Global Challenge* (2009), and called for a global ban on plastic bags to save oceans.

Current market transformation phase: phase 4

In 2007, the Californian city of San Francisco became the first U.S. city to ban plastic bags. Since San Francisco's trend-setting ban in 2007, plastic bags have been the subject of a mishmash of regulation in California: 90 cities and counties across the state now have some form of plastic bag ban—and more are on the way. And this is not only happening in the U.S. Countries such as South Africa, China, Taiwan, and Macedonia have implemented a total ban on the lightweight plastic bag.[356] And, in

November 2013, the European Commission published a proposal aiming to reduce the use of lightweight plastic (less than 50 microns thick) carrier bags.[357] Under the proposal, EU Member States can choose the most appropriate measures to discourage the use of plastic bags.

Future scenario

We are not there yet, the plastic garbage patch is still growing and recently another "plastic soup" location was discovered in the North Atlantic Gyre. But, thanks to Moore, our "use and throw away" mentality toward plastic has become more of an issue than ever. Countries, cities, and supermarkets are starting to ban lightweight plastic bags. And that's a strong beginning.

***

Topic	Name (organization)	Location	Period
Counterfeit medicines	Bright Simons (mPedigree)	Ghana	2007 to present

Link to tough problem

Counterfeit medicines are a serious problem, particularly in developing counties where law enforcement is weak and a solid regulatory framework is often lacking. In recent years the problem has become worse, as globalization and the rise of the internet have resulted in a greater consumption of counterfeit medicines. The World Health Organization estimated that one out of four medicines sold in a street market in developing countries is counterfeit. About 2,000 people die every day due to counterfeit medicines.[358]

Action to make a change

In 2007, Bright Simons and his team rolled out a pilot in Ghana in which mobile phone technology was used to determine whether packaged medicines had been properly tested and certified. According to Simons:

> You can send a free text message and get a reply in a few seconds verifying that a medicine or chemical is authentic. In

addition, distributors and other middlemen can check the codes to verify that the supply has not been compromised. This helped reveal to a major Indian company that there was pilfering at a depot. Genuine antimalarial medicines would be replaced by counterfeits. The shady characters cannot get away with this anymore. If we had not stopped these leakages in the supply chain, they could have put thousands of patients at risk.[359]

Current market transformation phase: phase 2 moving to phase 3

In 2008, Nigerian health officials showed interest in replicating the concept there. In 2009, mPedigree was launched as a business, as Simons and his team realized that a majority of the companies didn't want to do business with NGOs. The system is currently used in Ghana, Nigeria (where its codes are on 50 million packages of antimalarial drugs alone), Kenya, and India, with pilots in Uganda, Tanzania, South Africa, and Bangladesh.[360] Relationships have been established with a growing number of pharmaceutical companies, NGOs, and government agencies.[361]

Future scenario

Next to an expansion into other countries, mPedigree is currently also expanding to seeds, cosmetics, and other products. New technical applications are also emerging that offer many opportunities to improve the traceability of products. According to Simons: "We have built a major platform for supply chains in the developing world."[362]

***

Topic	Name (organization)	Location	Period
Sustainable banking in the Netherlands	Various organizations	The Netherlands	2007 to present

Link to tough problem

Since the financial crisis started in 2007 public opinion about financial institutions has changed.[363,364] Instead of banks being driven by short-term self-interest while externalizing their costs on others, a growing number of people want stronger financial monitoring of financial institutions and more behavior driven by long-term considerations.[365]

Action to make a change

High bonuses, short-term benefits, and risky or unethical investments have attracted considerably more attention in the media than previously. Numerous campaigns have been launched all around the world, of which the Occupy Movement is probably the most well known.[366] At the same time a couple of financial institutions in the Netherlands started to fit their products to this changing demand. By focusing more on transparency, investing in socially and environmentally friendly products, and reducing bonuses, these first movers aimed to cater to public opinion, while at the same time increasing their own market share.

Current market transformation phase: phases 1 and 2

It's hard to say whether the Dutch financial sector is now in the first or the second phase. On the one hand, it can be argued that it's in the first mover and competition phase. The two most well-known first movers are Triodos Bank (founded in 1980) and ASN Bank (part of SNS Reaal, founded in 1960). Although these two banks were founded quite some time ago, more recently they have started to grow rapidly in terms of customer base and managed assets.[367] Moreover, conventional banks started to offer "sustainable" savings accounts.[368] On the other hand, in 2012 the market share in payment and savings deposits of Triodos and ASN Bank together was just 1.8%. Therefore, it can be argued that everything is still more or less the same with the sector still in the awareness and project phase.

Future scenario

Unless we find ways for banks to start competing on issues such as sustainable and responsible banking, the sector will remain stuck in its current phase. Mainstream, first-mover banks must be able to agree on a code of conduct and use that code in their marketing to differentiate themselves from their competitors on issues such as transparency, low/no bonuses, no speculation, responsible investments, and client focus.

Otherwise the sector will continue to compete on the wrong issues—the next crisis is already in the making.

***

Topic	Name (organization)	Location	Period
Employment discrimination	Lilly Ledbetter	United States	1998 to present

Link to tough problem

In 1963 John F. Kennedy signed the U.S. Equal Pay Act into law, which came into force in 1964. This law aimed to close the wage gap between men and women; at that time women with full-time jobs earned on average between 59% and 64% of their male counterparts for doing exactly the same work. Now, 50 years and a number of other laws later, the wage gap has narrowed but is still significant. In 2012, on average, women earned 80.9% of men's wages.[369]

Action to make a change

From 1979 till 1998 Lilly Ledbetter worked for Goodyear in the U.S. Shortly before she retired, she received an anonymous note which stated that she was earning significantly less money than her male counterparts. She sued Goodyear, accusing them of gender discrimination, but then lost on appeal.[370] Over the next eight years her case made it to the Supreme Court, but "the Supreme Court voted 5–4 that Ledbetter was not entitled to compensation because she filed her claim more than 180 days after receiving her first discriminatory paycheck."[371] She started an intensive lobbying campaign and her cause was eventually taken up by Barack and Michelle Obama.[372]

Current market transformation phase: phase 4, going back to phase 1 again

On January 29, 2009 the Lilly Ledbetter Fair Pay Act was signed into law. This new law resets the 180 day limit for lawsuit regarding pay discrimination with each new paycheck. Before, as can be seen in the Supreme Court's ruling, it was only possible to sue a company within 180 days after the first payment.[373] Now, more than five years after Obama signed the Act, its impact has been relatively limited. The wage gap still exists and may even have widened since the law came into force, probably because of the economic crisis.[374] Nevertheless, Ledbetter is widely praised for her efforts on gender equality, and nowadays she travels across the U.S. urging minorities to claim their civil rights.[375]

Future scenario

Pay discrimination appears to be deeply rooted in American culture. In order to change this, a long-term strategy is needed, bridging the wage gap step by step. Transparency and benchmarking will be important factors to solving this. Eventually, it has to become attractive and competitive for employers to pay equal wages before we can reach a level playing field situation.

***

Topic	Name (organization)	Location	Period
Agricultural finance	Lucas Simons (SCOPEinsight) (disclaimer: the author of this book!)	The Netherlands and worldwide	2010 to present

Link to tough problem

Global agriculture is one of the most unsustainable sectors in the world. In order to feed 10 billion people, the agricultural sector needs to become more professional to increase productivity in a sustainable way. To become more professional, farmers need access to finance, inputs, services, and markets to develop professional and sustainable practices. SME farmer producer organizations have the greatest need for investments and offer the largest opportunity for business partners such as banks, traders, and investment funds. However, financial institutions, service providers, and suppliers are often not interested in doing business with this segment; they consider it to be risky, non-transparent, and an unattractive business case overall.

Action to make a change

In 2010, SCOPEinsight, a social enterprise founded in the Netherlands by Lucas Simons (author of this book), developed the first rating methodology to assess farmer organizations on their level of professionalism and economic viability.[376] Assessing farmer organizations on their business capabilities has two benefits:

1. The farmer organizations acquire improved insight into their strengths and weaknesses, and understand why and how they should improve themselves to become a better business, in cooperation with the capacity-builders that are active.
2. The better farmer organizations receive a credit rating report and they can be linked to banks, funds, and input providers who are interested in doing business with the better farmers. This simple intervention reduces costs, risks, connects supply and demand, and creates the conditions for markets to function more efficiently.

Current market transformation phase: phase 1 going to phase 2

After more than four years of operation, SCOPEinsight has rated over 350 producer organizations in more than ten countries. It is not only helping cooperatives and farmer organizations to become more professional and run their farms as a business, but is also linking the better ones to business partners. It is currently working with a growing number of banks, traders, and NGOs in various countries to implement and improve its rating methodology as a common language to link them all together. In 2013, the Dutch Embassy in Kenya selected SCOPEinsight, together with partner organizations, to establish a large program to roll out these services. In early 2014, the ABN AMRO social investment fund purchased an equity stake in the company, giving it not only sufficient liquidity to invest and grow as a business, but also more international standing and recognition that its methodologies are sound and robust. Even more recently, the International Finance Corporation (IFC), part of the World Bank group, has recognized its methodology and is co-sponsoring, co-branding, and co-implementing the methodology in its programs. Currently, the methodology is attracting different first movers like development banks, traders, capacity builders, and large businesses, all looking for better producer organizations to do business with.

The initial success of SCOPEinsight is triggering other initiatives to come up with similar concepts. The second phase of market transformation is about to begin in agricultural finance.

For his ability to bring systemic change through SCOPEinsight Lucas Simons was recognized as an Ashoka Fellow in 2013.[377]

Future scenario

The second phase of market transformation will endure for a few more years. It is a major challenge to create a mechanism that will enable first-mover banks, funds, input suppliers, and service providers to compete on doing business with farmer organizations. If this phase is done successfully, competition will come and compete on similar concepts and value propositions. This would be a welcome sight as it is proof that the sector is maturing. The step after that will be to formalize the investment and business landscape, to develop a common vision and roadmaps, and make more formal arrangements between banks, raters, capacity builders, and governments to work towards that goal.

***

Appendix 1: Sector fact sheets of coffee, cocoa, and palm oil

SUSTAINABILITY RELATED FACTS AND FIGURES OF THE COFFEE SECTOR

COFFEE PRODUCING AND CONSUMING COUNTRIES

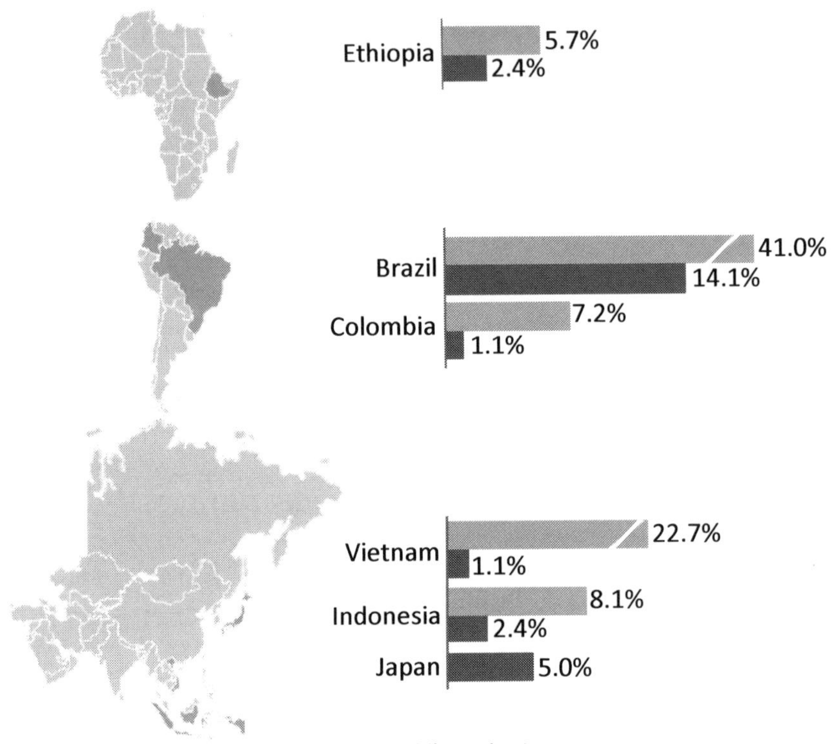

Top coffee producing countries, % of the world's production.
Top coffee consuming countries, % of the world's consumption.

COFFEE VALUE CHAIN

Characteristics: the majority of coffee producers are small-scale farmers, doing primary coffee processing (drying, hulling, etc.) by themselves. A high number of intermediaries challenges the transparency and traceability in the coffee sector.

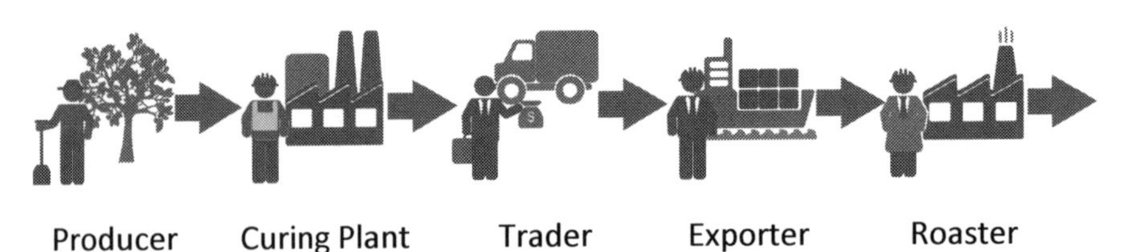

Europe 40.0%

USA 15.9%

Source: ICO, 2012

SECTOR FACTS

- There are around 25 million coffee producers in the world. A majority of them are smallholders dependent on coffee for their livelihood.[1]
- Coffee is the most widely traded commodity worldwide after petroleum.[2]
- Four single roaster companies buy more than 50% of all of the annual production. It is believed to have been a major factor in the plunge in green coffee prices in 2001.[3]

MAIN SUSTAINABILITY ISSUES

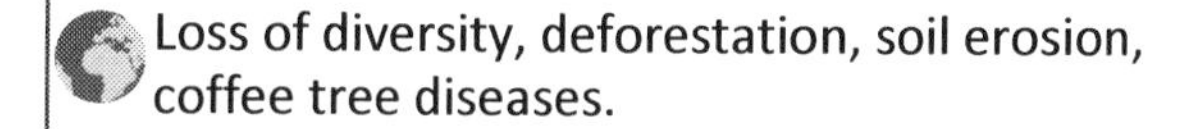

Loss of diversity, deforestation, soil erosion, coffee tree diseases.

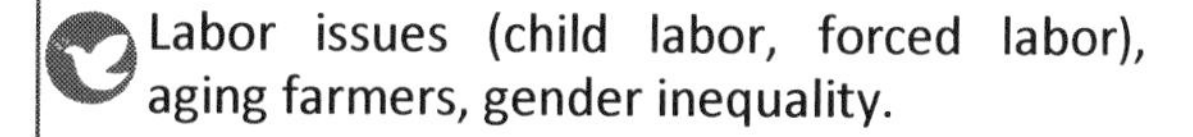

Labor issues (child labor, forced labor), aging farmers, gender inequality.

Low productivity, low wages, low level of farmer organization, coffee tree diseases and aging trees.

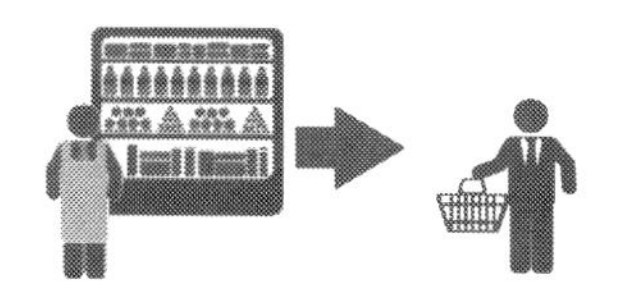

Sources: 1. Coffee Barometer, 2012. 2. ICO. 3. www.coffeefacts.com, 2011.

MAIN SUSTAINABILITY STANDARDS

Organic: Creates a sustainable agriculture system producing food in harmony with nature, supports biodiversity, and enhances soil health.

Fairtrade (1989): Supports a better life for farmers through fair prices, community development, and environmental stewardship.

Rainforest Alliance (1995): Integrates biodiversity conservation, community development, rights of the workers, and good agricultural practices.

UTZ (2002): Sets basic social and environmental criteria for good agricultural and management practices.

4C (2007): Sets the baseline standards for the path to sustainable production.

TRANSFORMATION CURVE OF THE COFFEE SECTOR

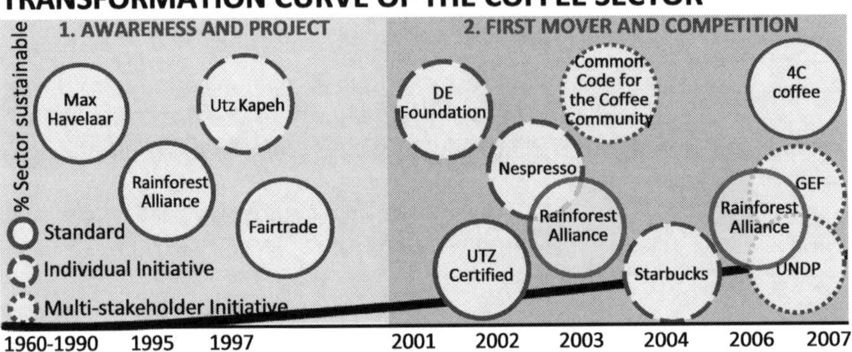

1962: The first International Coffee Agreement (ICA).

1995: Rainforest Alliance certifies the first coffee farm.

2001: ***Global coffee crisis. Coffee prices plummet to all-time low.***

2001: Oxfam publishes report "Bitter Coffee: How the Poor Are Paying for the Slump in Coffee Prices."

2004: Starbucks starts own private guidelines for quality and sustainable coffee production, Coffee and Farmer Equity Practices (CAFE) practices.

1988: ***ICA members failed to reach a new agreement on export quotas, causing the ICA to break down.***

1989: ***First Max Havelaar fairtrade coffee sold in supermarkets.***

1997: Fairtrade Labelling Organization established as labeling and harmonization umbrella of all the various initiatives.

1997: ***Utz Kapeh is set up as a foundation in Guatemala by Ahold.***

2006: Rainforest Alliance launches a biodiversity coffee conservation project, supported by the UNDP with a Global Environment Facility (GEF) grant.

Late 1990s: ***Vietnam, supported by the World Bank, drastically increases production of coffee from 1.5 million to 15 million bags in ten years.***

2002: Oxfam Novib targets major four coffee roasters in "Make it Fair" campaign and issues report "Mugged: Poverty in Your Coffee Cup."

2002: Utz Kapeh became an independent foundation and certified the first coffee farm.

2002: DE Foundation established to improve the living conditions of small coffee and tea farmers.

2003: Nespresso partners with Rainforest Alliance to implement its AAA Sustainable Quality Program guidelines.

2003: ***The Coffee Coalition begins campaign to make coffee roasters accept their responsibility for the conditions in the coffee sector.***

2003: ***The Common Code for the Coffee Community (4C) is set up – a platform of trade unions and social movements fighting for better working and living conditions.***

2007: 4C coffee is available in the market.

2007: Government of Minas Gerais, the largest coffee producing state in Brazil, launch Certifica Minas Café.

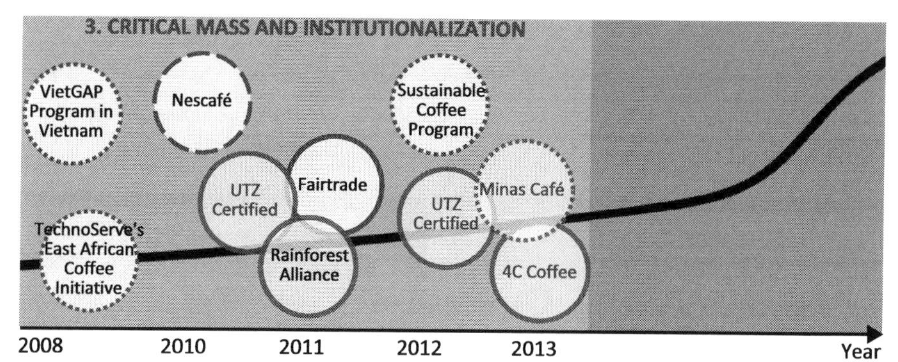

2008: Vietnamese government starts VietGAP

2008: TechnoServe's East African Coffee Initiative is started.

2008: Tchibo publishes its first sustainability report.

2010: Dutch coffee industry sets target to increase the total share of sustainably produced coffee to 75% by 2015.

2010: Nestlé launches the Nescafé Plan, bringing under one umbrella Nestlé's commitments on coffee farming, production, and consumption.

2011: Sara Lee, Starbucks, and Tchibo set public CSR targets.

2011: Fairtrade, Rainforest Alliance, and UTZ Certified release joint statement and commit to work together.

2012: ***Nestlé, Kraft, Tchibo, and DE start the pre-competitive sustainability program through IDH (Sustainable Coffee Program).***

2012: Kraft Foods changes its name to Mondelēz International and starts "Coffee Made Happy" program.

2013: Production challenges: depleted stock and coffee rust disease in Central America.

2013: UTZ Certified and Minas Café consolidate a long-term commitment.

2013: ***High turnout of industry, trade, NGOs, and producers at 4C's vision 2020 session shows commitment of the sector to developing a shared vision.***

2013: The 4C Association has formalized a technical cooperation agreement with Minas Café.

16% of available coffee was certified by the standard systems[1]

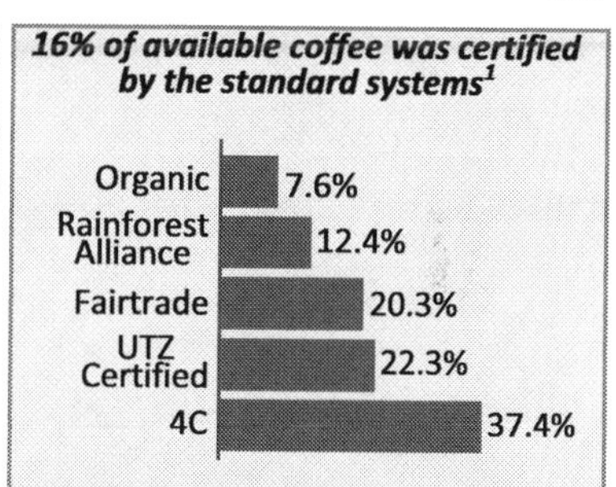

THE NEXT STEPS FOR THE COFFEE SECTOR:

- Embrace an overarching vision for a sustainable and vital coffee sector.
- Make a clear distinction between competitive and non-competitive.
- Prepare the strategy at the operational level.
- New industry standard will be translated into common industry standard.
- Prepare for thorough discussion on who are the coffee farmers of the future.
- Reach out to national governments to align agendas.
- Prepare for massive co-investments in the sector.
- Lobby for government support and tax breaks for sustainable coffee.

1. 2010.
Source: TCC, 2012. Coffee Barometer.

SUSTAINABILITY RELATED FACTS AND FIGURES OF THE COCOA SECTOR

COCOA PRODUCING AND IMPORTING COUNTRIES

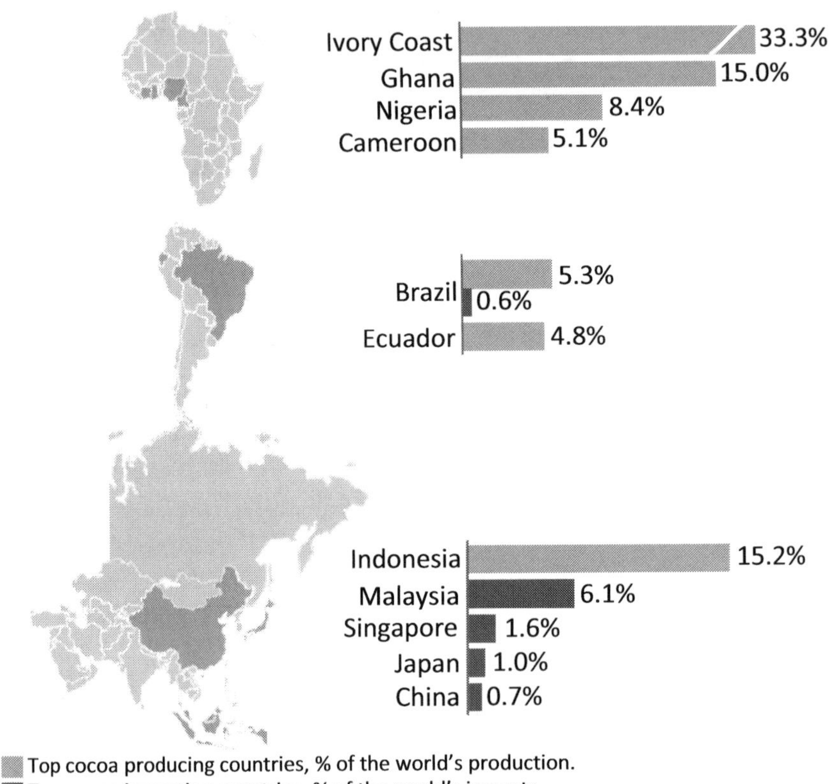

Top cocoa producing countries, % of the world's production.
Top cocoa importing countries, % of the world's imports.

COCOA VALUE CHAIN

Characteristics: farmers often are illiterate and unorganized. The cocoa supply chain is long and disorganized, with a complex trading network including a large number of intermediaries. It makes the transparency and traceability challenging.

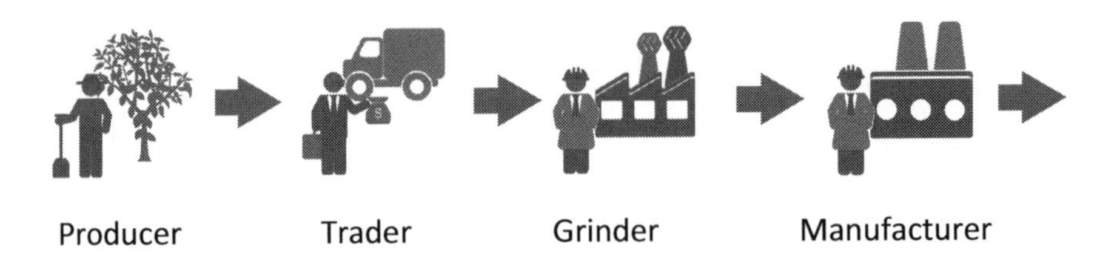

SECTOR FACTS

- >90% of the world's cocoa is grown on 5.5 million small farms.[1]
- The average income of West African cocoa farmers and their dependents is far below the level of absolute poverty (US$1–1.25 day). Insufficient income and a shortage of workers in rural areas force the families of the farmers to work excessive hours.[2]
- A substantial shortfall of cocoa is expected in 2020 due to the growing demand and poor agricultural practices. The expected demand is 4.5 MT while 3.5 MT of cocoa beans are produced each year.[3]

Europe

37.9%

N. America

10.1%

Source: FAOSTAT, 2011

MAIN SUSTAINABILITY ISSUES

Deforestation, biodiversity loss, soil depletion, water pollution.

Human rights violations (child and forced labor), widespread poverty, illiteracy, unsafe working conditions, aging farmer population.

Volatile prices due to speculation, inadequate bargaining power of smallholders, poor quality, low yields, lack of traceability, inadequate infrastructure, plant diseases.

MAIN SUSTAINABILITY STANDARDS

Organic: Creates a sustainable agriculture system that produces food in harmony with nature, supports biodiversity, and enhances soil health.

Fairtrade (1993): Supports a better life for farmers through fair prices, community development, and environmental stewardship.

Rainforest Alliance (1997): Integrates biodiversity conservation, community development, rights of the workers, and good agricultural practices.

UTZ (2007): Sets basic social and environmental criteria for good agricultural and management practices.

CEN (in development): Sets standards for production, traceability, and conformity.

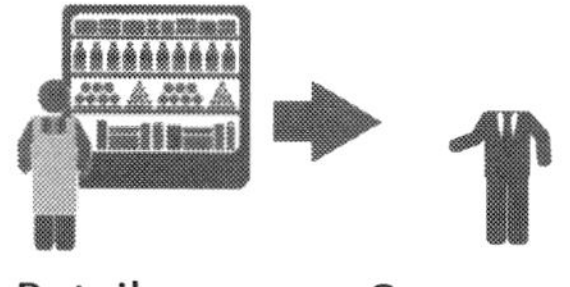

Retailer Consumer

Sources: 1. Seas of Change, 2012. 2. Cocoa Barometer, 2012; World Bank. 3. World Cocoa Foundation, 2012.

TRANSFORMATION CURVE OF THE COCOA SECTOR

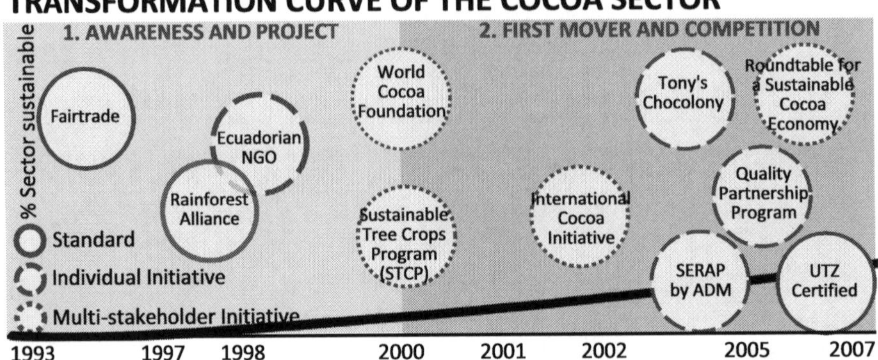

1990s: Witch's Broom disease in Latin America.

1993: Max Havelaar (later Fairtrade) sold first fairtrade chocolate in Europe. One year later it entered the UK market.

1997: Rainforest Alliance cocoa starts program with Ecuadorian conservation group.

1998: ***Prominent TV documentaries and reports in newspapers on child slavery.***

1998: Chocolate Day (Divine Chocolate now) is set up.

2000: ***World Cocoa Foundation is established with a mission to encourage sustainable cocoa practices.***

2000: ITTA launches Sustainable Tree Crops Program (STCP) with the aim of generating growth in rural income among tree crop farmers in West/Central Africa.

2001: ***Harkin–Engel protocol mandates the US federal system to force the chocolate industry to end slavery in its supply chain by 2005.***

2001: Industry gets fully engaged in an effort to address the issues in sector by investing in STCP.

2002: ***International Cocoa Initiative (ICI) is set up to oversee the elimination of child and forced labor.***

2005: Tony's Chocolony "slave free chocolate" launched.

2005: Harkin–Engel extended till 2008.

2005: Barry Callebout starts its Quality Partnership Program.

2005: Nestlé partners with Rainforest Alliance.

2005: ADM starts Socially and Environmentally Responsible Agricultural Practices (SERAP) program.

2005: Cadbury buys Green & Black's, organic chocolate.

2005: International Labor Rights Fund files suit on behalf of former child laborers against Nestlé, Archer Daniels Midland, and Cargill.

2007: ***Major companies join UTZ Good Inside Cocoa Program.***

2007: First Roundtable for a Sustainable Cocoa Economy.

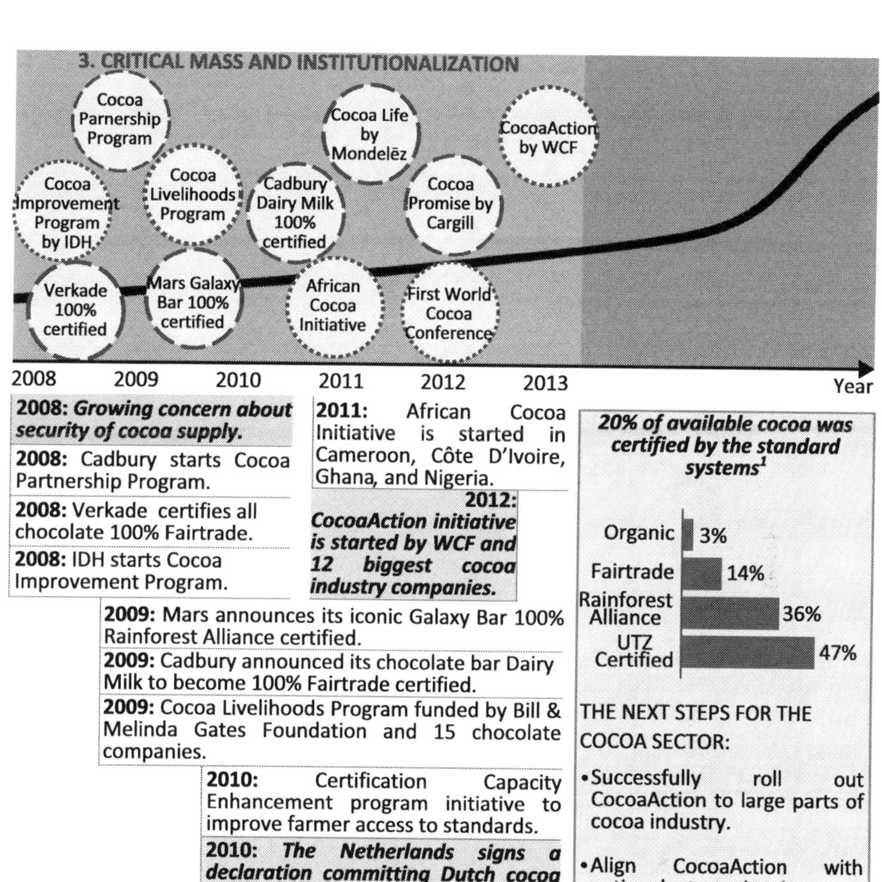

2008: *Growing concern about security of cocoa supply.*

2008: Cadbury starts Cocoa Partnership Program.

2008: Verkade certifies all chocolate 100% Fairtrade.

2008: IDH starts Cocoa Improvement Program.

2011: African Cocoa Initiative is started in Cameroon, Côte D'Ivoire, Ghana, and Nigeria.

2012: *CocoaAction initiative is started by WCF and 12 biggest cocoa industry companies.*

2009: Mars announces its iconic Galaxy Bar 100% Rainforest Alliance certified.

2009: Cadbury announced its chocolate bar Dairy Milk to become 100% Fairtrade certified.

2009: Cocoa Livelihoods Program funded by Bill & Melinda Gates Foundation and 15 chocolate companies.

2010: Certification Capacity Enhancement program initiative to improve farmer access to standards.

2010: *The Netherlands signs a declaration committing Dutch cocoa sector to use 100% sustainable cocoa by 2025.*

2012: The first World Cocoa Conference. Abidjan Cocoa Declaration is signed.

2012: Sustainable Cocoa Forum launched in Germany to promote sustainable cocoa farming.

2012: Cocoa-MAP (Measuring and Progress) is launched for tracking the results in sustainable cocoa.

2012: European Committee for Standardization (CEN) aims to set up a standard that would harmonize private standards.

2012: Hershey's certifies 100% of Bliss chocolate.

2012: Oxfam Novib highlights poor working conditions of women in cocoa.

2012: Cargill launches Cocoa Promise program.

2012: Mondelēz launches Cocoa Life ten-year sustainability program of $400 million.

20% of available cocoa was certified by the standard systems[1]

Organic	3%
Fairtrade	14%
Rainforest Alliance	36%
UTZ Certified	47%

THE NEXT STEPS FOR THE COCOA SECTOR:

- Successfully roll out CocoaAction to large parts of cocoa industry.
- Align CocoaAction with national strategies in cocoa producing countries.
- Lobby for government support and tax breaks for sustainable cocoa.
- New industry standard will be translated into common industry standard.
- Prepare for a thorough discussion on what is a cocoa farmer of the future, what is not and what are the implications.

1. 2012.

Sources: UTZ Certified, 2012. Annual Report; RA website; ICCO data, 2011.

SUSTAINABILITY RELATED FACTS AND FIGURES OF THE PALM OIL SECTOR

PALM OIL PRODUCING AND CONSUMING COUNTRIES

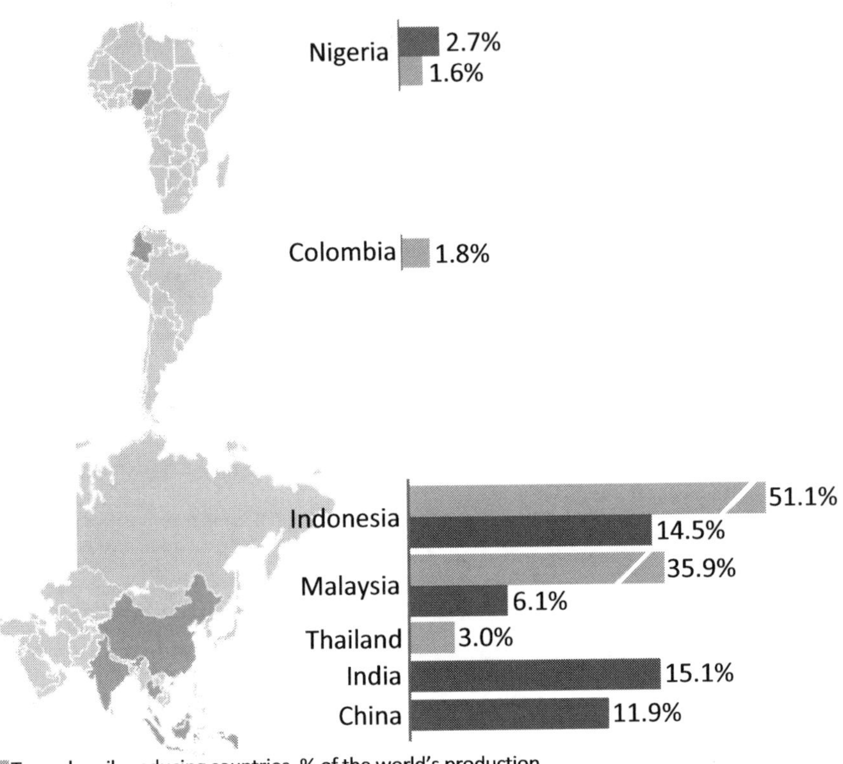

Top palm oil producing countries, % of the world's production.
Top palm oil consuming countries, % of the world's consumption.

PALM OIL VALUE CHAIN

Characteristics: complex and lacking transparency. Palm oil often is a hidden component in many products. In the case of the small farms, supplies from different sources are mixed at multiple stages, making the transparency and traceability very challenging. For the large plantations, traceability is easier.

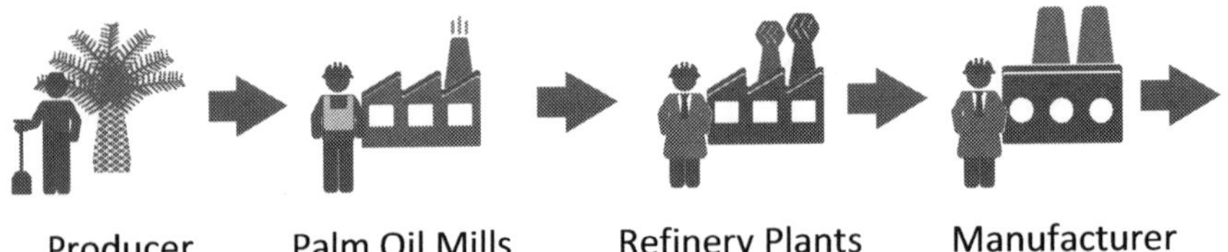

Europe

USA

Source: AOCS Lipid Library, 2012

SECTOR FACTS

- Palm oil demand is expected to double by 2020.[1]
- Palm oil is found in 50% of the products in supermarkets. Edible oil is derived from the pulp of the oil palm fruits and is used in foods (margarine, cooking oil, baked goods), consumer care products (creams, shampoos, make-up), and biofuels.[2]
- Relatively cheap, high-yielding (the yield is four to ten times higher than other oil crops), and is the most widely produced vegetable oil worldwide.[3]

MAIN SUSTAINABILITY ISSUES

Deforestation (clearing tropical forests and national parks), biodiversity loss, climate change, greenhouse gas emissions, water pollution.

Abuse of indigenous peoples' rights, appropriation of their land, poor labor conditions.

High levels of debt for smallholders, weak bargaining power, land conflicts.

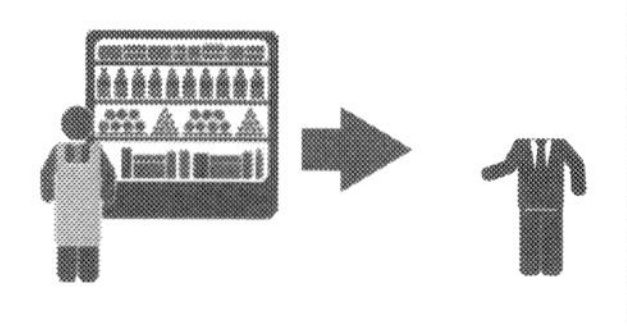

Retailer Consumer

Sources: 1. WWF, 2013.
2. Rainforest Alliance.
3. RSPO, 2012.

MAIN SUSTAINABILITY STANDARDS

Roundtable on Sustainable Palm Oil (2008): Ensures that food products and oleo-chemicals are produced without undue harm to the environment and society, and the volumes are traceable.

Indonesian Roundtable for Sustainable Palm Oil (2009): Based on existing Indonesian legislation, designed to ensure that all Indonesian oil palm growers conform to higher agricultural standards. The first national standard of this kind.

International Sustainability and Carbon Certification (2010): The certification scheme for palm oil used as a feedstock for biofuels under the EU-RED. It includes a carbon accounting mechanism.

TRANSFORMATION CURVE OF THE PALM OIL SECTOR

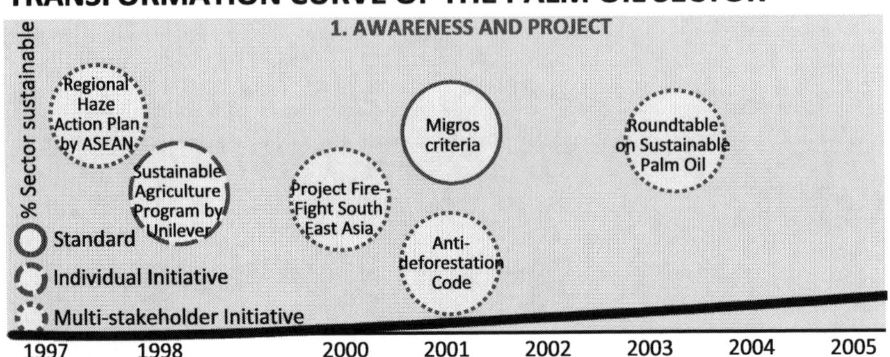

1965–1990: Rapid growth of palm oil production.

1997: Indonesia on fire: regional haze affects South-East Asia.

1997: ASEAN countries develop a Regional Haze Action Plan.

1997: Greenpeace research shows fires in Indonesia were deliberately set to clear land for plantations.

1997: WWF issued a milestone report: "The year the world caught fire."

1998: NGOs start public campaigns to raise awareness about cause of fires. New projects are launched such as SAWIT Watch.

1998: Unilever launches Sustainable Agriculture Programme.

1998: Indonesia pledges to implement "zero burning policy."

2000: WWF and IUCN implement Project Fire-Fight South-East Asia.

2000: Greenpeace's report "Funding Forest Destruction" identifies Dutch banks as major investors in palm oil plantations.

2000: Dutch financial institutions launch public CSR programs.

2000: WWF and Stichting Doen start to showcase sustainable palm oil.

2000: Unilever partners with WWF to develop economic, environmental, and social criteria for sustainable palm oil production.

2001: Dutch financial institutions sign "Anti-deforestation Code."

2001: Migros develops criteria for sustainable palm oil production assisted by partners Proforest and WWF.

2002: Kalimantan on fire: hotspots are plantations.

2002: WWF invites 30 retailers, food manufactures, processors, and traders to the first multi-stakeholder meeting in London.

2003: Multi-stakeholder initiative is formalized in Roundtable on Sustainable Palm Oil (RSPO).

2003: RSPO develops and adopts principles and criteria for sustainable palm oil.

2005: Friends of the Earth report "The Oil for Ape Scandal" published; renewed NGO campaigns follow.

2. FIRST MOVER AND COMPETITION

2007: Malaysian Palm Oil Council (MPOC) launches a fund to improve sustainable practices and biodiversity conservation.

2008: First RSPO Certified Sustainable Palm Oil.

2007: RSPO contracts UTZ Certified as its provider for traceability services.

2009: Greenpeace campaign "Unilever: Don't Destroy Forests."

2009: Indonesian Sustainable Palm Oil standard launched (ISPO).

2010: International Sustainability and Carbon Certification (ISCC) developed with support from German government.

2010: ***Unilever announces it is to source 100% sustainable palm oil by 2015.***

2010: ***Dutch government pledges to import 100% sustainable palm oil by 2015.***

2011: Indonesian government implements a moratorium on conversion of natural, primary, and peat forests.

2013: After a period of drought, large fires return.

2013: RSPO receives criticism as members are implicated in the forest fires.

2013: ***Unilever announces that, by the end of 2014, it will buy all palm oil from traceable sources.***

2013: GAR launches forest conservation project after pressure from NGOs.

2013: Indonesia becomes final member of South-East Asian antipollution pact.

2013: Cooperation between ISPO and RSPO toward sustainable palm oil in Indonesia.

2013: Blommer, Starbucks, and Ferrero follow with CSR targets.

14% of produced palm oil is RSPO certified[1]

THE NEXT STEPS FOR THE PALM OIL SECTOR:

- Develop and embrace an overarching global vision on what a sustainable and vital palm oil sector means.
- Palm oil industry moves beyond certification as the main instrument.
- Lobby for government support and tax breaks for sustainable palm oil.
- Make a clear distinction between the roles of industry and government.
- Give operational meaning to this strategy. A new standard will be the translation of this strategy.
- Prepare for a thorough discussion on what is a palm oil farmer of the future, what is not and what are the implications.
- Reach out to national governments to align agendas.

1. 2013. The figures of other certification schemes (ISCC, ISPO) are not publicly available.
Source: sustainablepalmoil.org.

Appendix 2: Sources

Editorial board (in alphabetical order)

Chattopadhayah, Shatadru	Managing Director, Solidaridad South and South-East Asia
Dijk, Gerda van	Director of the board at the Zijlstra Center for Public Control and Governance and Professor of Organizational Ecology at University of Tilburg
Kissmann, Edna	Co-owner and Senior Partner in Kissmann Langford
Oldenburg, Felix	Director Ashoka Germany and Europe
Peppelenbos, Lucian	Director of Learning and Innovation at IDH
Schoemaker, Anne	Management Team, SCOPEinsight

Interview respondents (in alphabetical order)

Name	Function (date of interview)
Bexell, Anna	Global Supply Chain Advisor, UNDP (July 24, 2013)
Bocklandt, Nick	Founder of Utz Kapeh (December 20, 2012)
Boetekees, Gemma	Global Network Director, FSC International (August 29, 2013)
Bos, Bernedine	Manager Government and Sectors, CSR Netherlands (October 10, 2013)
Browning, David	Senior Vice President Strategic Initiatives, TechnoServe (August 22, 2013)
Cameron, Rob	Executive Director, SustainAbility; former CEO, Fairtrade International (October 8, 2013)
Chattopadhayay, Shatadru	Managing Director, Solidaridad South & South-East Asia (August 12, 2013)
Child, Alastair	Cocoa Sustainability Director, Mars (August 28, 2013)
Debenham, Nicko	Former Head of Cocoa, Armajaro; currently Head of Sustainability, Barry Callebaut (July 22, 2013)
Dros, Jan Maarten	International Palm Oil Program Coordinator, Solidaridad (July 8, 2013)
Duffey, Tracey	Funding Manager, Source Trust (August 15, 2013)
Ferrigno, Simon	Consultant on sustainable and organic farm systems (August 7, 2013)
Goodall, Nick	Former CEO, Bonsucro (August 6, 2013)
Groot, Han de	Executive Director, UTZ Certified (June 25, 2013)
Guémas, Matthieu	Former Corporate Responsibility Manager, Cargill; currently Head of Business Development for Europe, Africa and Asia at GeoTraceability (June 17, 2013)

Guyton, Bill	President, World Cocoa Foundation (July 12, 2013)
Holzman, Caren	Director, Enabling Outcomes Ltd (July 30, 2013)
Kahn, Hammad Naqi	Global Cotton Leader, WWF International (August 1, 2013)
Klerk, Paul de	Coordinator Economic Justice Program, Friends of the Earth Europe (March 14, 2014)
Kreider, Karin	Executive Director, ISEAL Alliance (July 17, 2013)
Kuhrt, Cornel	Coffee industry expert (September 9, 2013)
Lelijveld, Daudi	Former Vice President Cocoa Sustainability, Barry Callebaut (August 8, 2013)
Mascotena, Agustín	Executive Director, Responsible Soy (July 31, 2013)
Mechielsen, Frank	Policy advisor, Oxfam Novib (June 12, 2013)
Melvin, Lise	Former CEO, Better Cotton Initiative (August 19, 2013)
Mensink, Janet	International Program Coordinator Sustainable Cotton & Textiles, Solidaridad (July 8, 2013)
Miltenburg, Stefanie	Director International Corporate Social Responsibility, D.E Master Blenders 1753 and Director, Douwe Egberts Foundation (June 21, 2013)
Oorthuizen, Joost	Executive Director, IDH (November 12, 2013)
Pedersen, Leif	Senior Commodities Advisor, UNDP (August 2, 2013)
Roozen, Nico	Managing Director, Solidaridad Netherlands (July 2, 2013)
Rosenberg, David	Group Sustainability Officer, Ecom Agroindustrial Corp (August 23, 2013)

Rutten-Sülz, Melanie	Executive Director, 4C Association (August 5 and August 13, 2013)
Schaeffers, Loeki	Senior Program Manager International, CSR Netherlands (October 10, 2013)
Schmitz-Hoffmann, Carsten	Executive Director Private Sector Cooperation, GIZ (July 16, 2013)
Singh, Pramod	IKEA Cotton Leader (August 12, 2013)
Verburg, Esther	Former General Manager, MADE-BY Benelux (August 7, 2013)
Verburg, Johan	Senior Advisor Program Development & Private Sector Engagement in Agribusiness, Oxfam Novib (June 12, 2013)
Vernooij, Marcel	Management team member of the Sustainable Economic Development Department (DDE), Dutch Ministry of Foreign Affairs (June 17, 2013)
Vis, Jan Kees	Global Director Sustainable Sourcing Development, Unilever (September 12, 2013)
Wakker, Eric	Senior Consultant, Aidenvironment (March 2014; email)
Webber, Darrel	Secretary General, Roundtable on Sustainable Palm Oil (September 19, 2013)
Weert, Marian van	Specialist Private Sector Engagement, ICCO (October 7, 2013)
Wijn, Annemieke	Senior Director for Commodity Sustainability, Kraft Foods (August 5, 2013)
Wise, Bruce	Global Product Specialist, Sustainable Business Advisory, International Finance Corporation (July 11, 2013)
Ywema, Peter Erik	General Manager, Sustainable Agricultural Initiative Platform (November 21, 2013)
Züblin, Johann	Specialist on sustainable development (July 31, 2013)

Notes

Chapter 1: Guatemala, where it all began

1. Utz Kapeh changed its name to UTZ Certified in 2007 to accommodate certification of other crops such as cocoa, tea, and nuts.
2. G. Greenfield (2002) "Vietnam and the World Coffee Crisis: Local Coffee Riots in a Global Context," http://focusweb.org/publications/2002/Vietnam-and-the-world-coffee-crisis.html, accessed August 18, 2014.
3. U.S. Department of Agriculture Foreign Agricultural Service (2013) "Guatemala: Coffee Annual," http://gain.fas.usda.gov/Recent%20GAIN%20Publications/Coffee%20Annual_Guatemala%20City_Guatemala_5-24-2013.pdf, accessed August 25, 2014.
4. M. Sheridan (2013) "Coffee Rust and Farmworkers," *Catholic Relief Services*, http://coffeelands.crs.org/2013/05/coffee-rust-and-farmworkers, accessed May 9, 2014.
5. D. Lotter (2003) "The Price, Processing and Production Challenges of Growing Coffee Profitably and Sustainably in Guatemala," *Rodale Institute*, www.newfarm.org/international/guatemala/coffee.shtml, accessed May 9, 2014.
6. As explained in Chapter 2, when prices are high, hog farmers will invest more to benefit from the high prices. The effect, however, is delayed due to the breeding and growing time of the hogs. Because of a sudden oversupply of hogs after this delay, the market becomes saturated, which leads to a decline in prices. As a result of this, production is reduced, but it takes time again before this affects market prices. Then this leads once again to undersupply and increasing prices. This cycle repeats itself.
7. Interview with Nick Bocklandt, founder of Utz Kapeh, December 20, 2012.

Chapter 2: What you eat impacts the world

8. M.L. King (1967) "A Christmas Sermon on Peace," www.ecoflourish.com/Primers/education/ChristmasSermon.html, accessed May 8, 2014.
9. International Labour Organization, "Agriculture; Plantations; Other Rural Sectors," www.ilo.org/global/industries-and-sectors/agriculture-plantations-other-rural-sectors/lang--en/index.htm, accessed May 8, 2014.

10. Global Agriculture, "Industrial Agriculture and Small-scale Farming," www.globalagriculture.org/report-topics/industrial-agriculture-and-small-scale-farming.html, accessed May 8, 2014.
11. FAO (2013) "Resilient Livelihoods: Disaster Risk Reduction for Food and Nutrition Security," www.fao.org/docrep/015/i2540e/i2540e00.pdf, accessed August 26, 2014.
12. African Development Bank Group (2010) "Agriculture Sector Strategy 2010–2014," www.afdb.org/fileadmin/uploads/afdb/Documents/Policy-ocuments/Agriculture%20Sector%20Strategy%2010-14.pdf, accessed August 26, 2014.
13. USAID Feed The Future, "Kenya," www.feedthefuture.gov/country/kenya, accessed May 8, 2014.
14. FAO (2013) "FAO Statistical Yearbook 2013: World Food and Agriculture" (Rome: Food and Agriculture Organization of the United Nations).
15. Government of the Netherlands (2013) "Dutch Government Position on Scale of Intensive Livestock Production," www.government.nl/news/2013/06/14/dutch-government-position-on-scale-of-intensive-livestock-production.html, accessed May 8, 2014.
16. World Bank (2014) "World Development Indicators: Rural Environment and Land Use," http://wdi.worldbank.org/table/3.1, accessed May 8, 2014.
17. J.A. Foley (2011) "Can We Feed the World and Sustain the Planet? A Five-step Global Plan Could Double Food Production by 2050," *Scientific American*, November 2011: 60-65.
18. UNCCD, "Desertification, Land Degradation and Drought (DLDD): Some Global Facts and Figures," www.unccd.int/Lists/SiteDocumentLibrary/WDCD/DLDD%20Facts.pdf, accessed August 24, 2014.
19. IFAD, "Water Facts and Figures," www.ifad.org/english/water/key.htm, accessed May 8, 2014.
20. A.K. Chapagain and A.Y. Hoekstra (2007) "The Water Footprint of Coffee and Tea Consumption in the Netherlands," *Ecological Economics* 64: 109-18.
21. Water Footprint, "Product Water Footprints: Coffee & Tea," www.waterfootprint.org/?page=files/CoffeeTea, accessed May 8, 2014.
22. European Commission (2012) "The Awake Water Guide," www.generationawake.eu/en/consumption-guide/the-water-guide, accessed August 26, 2014.
23. IFAD, "Water Facts and Figures," www.ifad.org/english/water/key.htm, accessed May 8, 2014.
24. World Economic Forum (2009) "The Bubble Is Close to Bursting: A Forecast of the Main Economic and Geopolitical Water Issues Likely to Arise in the World During the Next Two Decades. Draft for Discussion at the World Economic Forum Annual Meeting 2009," www3.weforum.org/docs/WEF_ManagingFutureWater%20Needs_Discussion-Document_2009.pdf, accessed August 26, 2014.
25. NASA (2012) "Landsat Top Ten: A Shrinking Sea, Aral Sea," www.nasa.gov/mission_pages/landsat/news/40th-top10-aralsea.html, accessed May 8, 2014.
26. Nature Kenya, "Lake Naivasha," www.naturekenya.org/content/lake-naivasha, accessed May 8, 2014.
27. D.K. Dash (2013) "22 of India's 32 Big Cities Face Water Crisis," *The Times of India*, September 9, 2013, http://timesofindia.indiatimes.com/india/22-of-Indias-32-big-cities-face-water-crisis/articleshow/22426076.cms?referral=PM, accessed May 8, 2014.
28. UPI (2011) "Global Drop in Groundwater Levels Seen," www.upi.com/Science_News/2011/12/22/Global-drop-in-groundwater-levels-seen/UPI-60611324606874, accessed May 8, 2014.

29. FAO (2011) "Biodiversity for Food and Agriculture: Contributing to Food Security and Sustainability in a Changing World," www.fao.org/fileadmin/templates/biodiversity_paia/PAR-FAO-book_lr.pdf, accessed August 26, 2014.
30. S.M. Philpott, W.J. Arendt, I. Armbrecht, P. Bichier, T.V. Diestch, C. Gordon, R. Greenberg, I. Perfecto, R. Reynoso-Santos, L. Soto-Pinto, C. Tejeda-Cruz, G. Williams-Linera, J. Valenzuela, and J.M. Zolotoff (2008) "Biodiversity Loss in Latin American Coffee Landscapes: Review of the Evidence on Ants, Birds, and Trees," *Conservation Biology* 22.5: 1093-105.
31. E. Wakker (2000) *Funding Forest Destruction: The Involvement of Dutch Banks in the Financing of Oil Palm Plantations in Indonesia* (Greenpeace Netherlands).
32. NASA Earth Observatory, "Causes of Deforestation: Direct Causes," http://earthobservatory.nasa.gov/Features/Deforestation/deforestation_update3.php, accessed May 8, 2014.
33. PHYS.org (2013) "Amazon Deforestation Due in Part to Soybean Growing," http://phys.org/news/2013-09-amazon-deforestation-due-soybean.html, accessed May 8, 2014.
34. S. Wallace (2013) "Farming the Amazon," *National Geographic*, http://environment.nationalgeographic.com/environment/habitats/last-of-amazon, accessed August 26, 2014.
35. Greenpeace (2013) "Palm Oil Leading Cause of Indonesia Forest Destruction: RSPO Leaves Big Brands Exposed," www.greenpeace.org/international/en/press/releases/Palm-oil-leading-cause-of-Indonesia-forest-destruction-RSPO-leaves-big-brands-exposed, accessed May 8, 2014.
36. UNEP (2012) "Global Chemicals Outlook: Towards Sound Management of Chemicals," www.unep.org/pdf/GCO_Synthesis%20Report_CBDTIE_UNEP_September5_2012.pdf, accessed August 26, 2014.
37. Ecoworld (2011) "EPA Faces Pesticides, Endangered Species Lawsuit," www.ecoworld.com/animals/epa-faces-pesticides-endangered-species-lawsuit.html, accessed May 8, 2014.
38. C.A. Brühl, T. Schmidt, S. Pieper, and A. Alscher (2013) "Terrestrial Pesticide Exposure of Amphibians: An Underestimated Cause of Global Decline?" *Scientific Reports* 3: 1135, DOI: 10.1038/srep01135.
39. WWF, "Ganges River Dolphin," http://worldwildlife.org/species/ganges-river-dolphin, accessed May 8, 2014.
40. European Commission, "Descriptor 5: Eutrophication," http://ec.europa.eu/environment/marine/good-environmental-status/descriptor-5/index_en.htm, accessed August 24, 2014.
41. Nature, "Eutrophication: Causes, Consequences, and Controls in Aquatic Ecosystems," www.nature.com/scitable/knowledge/library/eutrophication-causes-consequences-and-controls-in-aquatic-102364466, accessed August 24, 2014
42. M.A. Beketov, B.J. Kefford, R.B. Schäfer, and M. Liess (2013) "Pesticides Reduce Regional Biodiversity of Stream Invertebrates," *Proceedings of the National Academy of Sciences of the United States of America* 110.27: 11039-43.
43. OUPblog (2006) "Carbon Neutral: Oxford Word of the Year," http://blog.oup.com/2006/11/carbon_neutral_, accessed May 8, 2014: "Being carbon neutral involves calculating your total climate-damaging carbon emissions, reducing them where possible, and then balancing your remaining emissions, often by purchasing a carbon offset: paying to plant new trees or investing in 'green' technologies such as solar and wind power." Erin McKean, Editor-in-Chief of the New Oxford American Dictionary 2e, said:

"The increasing use of the word carbon neutral reflects not just the greening of our culture, but the greening of our language. When you see first graders trying to make their classrooms carbon neutral, you know the word has become mainstream."

44. P. Thornton and L. Cramer (eds.) (2012) "Impacts of Climate Change On the Agricultural and Aquatic Systems and Natural Resources Within the CGIAR's Mandate," CCAFS Working Paper 23, *CGIAR Research Program on Climate Change, Agriculture and Food Security (CCAFS).*
45. Worldwatch Institute (2013) "Agriculture and Livestock Remain Major Sources of Greenhouse Gas Emissions," www.worldwatch.org/agriculture-and-livestock-remain-major-sources-greenhouse-gas-emissions-1, accessed May 8, 2014.
46. U.S. Environmental Protection Agency, "Global Greenhouse Gas Emissions Data," www.epa.gov/climatechange/ghgemissions/global.html, accessed May 8, 2014.
47. K. Hergoualc'h and L. Verchot (2012) "Changes in Soil CH_4 Fluxes from the Conversion of Tropical Peat Swamp Forests: A Meta-analysis," *Journal of Integrative Environmental Sciences* 9.2: 93-101.
48. FAO (2006) *Livestock's Long Shadow: Environmental Issues and Options* (Rome: Food and Agriculture Organization of the United Nations).
49. K.A. Johnson and D.E. Johnson (1995) "Methane Emissions from Cattle," *Journal of Animal Science* 73: 2483-92.
50. FAO (2006) *Livestock's Long Shadow: Environmental Issues and Options* (Rome: Food and Agriculture Organization of the United Nations).
51. Worldwatch Institute (2013) "Agriculture and Livestock Remain Major Sources of Greenhouse Gas Emissions," www.worldwatch.org/agriculture-and-livestock-remain-major-sources-greenhouse-gas-emissions-1, accessed May 8, 2014.
52. World Bank (2013) *Turn Down the Heat: Climate Extremes, Regional Impacts, and the Case for Resilience* (Washington, DC: World Bank): 18, 65, 105.
53. J. Vidal (2013) "Climate Change: How a Warming World is a Threat to Our Food Supplies," *The Guardian*, April 13, 2013, www.theguardian.com/environment/2013/apr/13/climate-change-threat-food-supplies, accessed May 8, 2014.
54. P. Thornton and L. Cramer (eds.) (2012) "Impacts of Climate Change on the Agricultural and Aquatic Systems and Natural Resources Within the CGIAR's Mandate," CCAFS Working Paper 23, *CGIAR Research Program on Climate Change, Agriculture and Food Security (CCAFS)*: 13.
55. J. Vidal (2013) "Climate Change: How a Warming World is a Threat to our Food Supplies," *The Guardian*, April 13, 2013, www.theguardian.com/environment/2013/apr/13/climate-change-threat-food-supplies, accessed May 8, 2014.
56. L. Geist (2013) "Rain Delays U.S. Planting, Topsoil Loss Could Cost Farmers Yield," *The Crop Site*, www.thecropsite.com/news/13732/rain-delays-us-planting-topsoil-loss-could-cost-farmers-yield, accessed May 8, 2014.
57. Met Office Hadley Centre (2012) *Climate Impacts on Food Security and Nutrition: A Review of Existing Knowledge*, www.metoffice.gov.uk/media/pdf/k/5/Climate_impacts_on_food_security_and_nutrition.pdf, accessed August 26, 2014.
58. IPCC (2007) *Climate Change 2007: Synthesis Report*, www.ipcc.ch/pdf/assessment-report/ar4/syr/ar4_syr.pdf, accessed August 26, 2014.
59. International Labour Office (2010) *Accelerating Action Against Child Labour*, report for the *International Labour Conference, 99th Session 2010.*
60. ILO (2012) "Forced Labour, Human Trafficking and Slavery," www.ilo.org/global/topics/forced-labour/lang--en/index.htm, accessed May 8, 2014.

61. FAO (2013) *The State of Food Insecurity in the World: The Multiple Dimensions of Food Security* (Rome: Food and Agriculture Organization of the United Nations).
62. IFAD (2011) *Rural Poverty Report 2011* (Rome: IFAD).
63. International Labour Office (2009) *Guide to the New Millennium Development Goals Employment Indicators* (Geneva: International Labour Office, Employment Sector).
64. ILO, "Hazardous Work," www.ilo.org/safework/areasofwork/hazardous-work/lang--en/index.htm, accessed May 8, 2014.
65. CDC, "Agricultural safety," www.cdc.gov/niosh/topics/aginjury/, accessed August 24, 2014.
66. ILO, "Child Labour in Agriculture," www.ilo.org/ipec/areas/Agriculture/lang--en/index.htm, accessed May 8, 2014.
67. ILO (2013) "Forced Labour: Facts and Figures," www.ilo.org/global/about-the-ilo/media-centre/issue-briefs/WCMS_207611/lang--en/index.htm, accessed May 8, 2014.
68. Global Slavery Index (2013) *The Global Slavery Index 2013*, www.globalslaveryindex.org, accessed August 26, 2014.
69. J. Campbell (2008) "A Growing Concern: Modern Slavery and Agricultural Production in Brazil and South Asia," *Human Rights & Human Welfare*, nn: 131-41.
70. Unicef, "The Time to Sow," www.unicef.org/pon00/leaguetos1.htm, accessed May 8, 2014.
71. World Food Programme (2014) "Who Are the Hungry?" www.wfp.org/hunger/who-are%20/, accessed May 8, 2014.
72. WHO (2014) "10 Facts on Obesity," www.who.int/features/factfiles/obesity/en/, accessed June 30, 2014.
73. UN Economic & Social Affairs (2013) *World Population Prospects: The 2012 Revision. Highlights and Advance Tables* (New York: United Nations).
74. C. McEvedy and R. Jones (1978) *Atlas of World Population History* (Harmondsworth, UK: Penguin).
75. United Nations, Department of Economic and Social Affairs, Population Division, Population Estimates and Projections Section, *World Population Prospects: The 2012 Revision* (New York: United Nations).
76. United Nations, Department of Economic and Social Affairs. Population Division, Population Estimates and Projections Section, *World Population Prospects: The 2012 Revision* (New York: United Nations).
77. J.-L. Nothias (2011) "Il naît chaque année plus d'enfants au Nigeria qu'en Europe," *Le Figaro*, August 17, 2011, www.lefigaro.fr/international/2011/08/17/01003-20110817ARTFIG00495-il-nait-chaque-annee-plus-d-enfants-au-nigeria-qu-en-europe.php, accessed June 10, 2014.
78. UN Economic & Social Affairs (2013) *World Population Prospects: The 2012 Revision. Highlights and Advance Tables* (New York: United Nations).
79. OECD (2012) *Economic Outlook 2012*, www.keepeek.com/Digital-Asset-Management/oecd/economics/oecd-economic-outlook-volume-2012-issue-2_eco_outlook-v2012-2-en#page3, accessed August 26, 2014.
80. C. Hanson (2013) *Food Security, Inclusive Growth, Sustainability, and the Post-2015 Development Agenda* (World Resources Institute).
81. T. Kastner, M.J.I. Rivas, W. Koch, and S. Nonhebel (2012) "Global Changes in Diets and the Consequences for Land Requirements for Food," *Proceedings of the National Academy of Sciences* 109.18: 6868-72.
82. L.B.J. Šebek and E.H.M. Temme (2009) "Human Protein Requirements and Protein Intake and the Conversion of Vegetable Protein into Animal Protein," *WUR, Rapport* 232, www.foodlog.nl/images/uploads/edepotlink_t4a421231_001.pdf, accessed August 26, 2014.

83. Neste Oil, "Demand for Biofuels Increasing," www.nesteoil.com/default.asp?path=1,41,537,2455,8529,12421, accessed May 8, 2014.
84. A. Demirbas (2008) "Biofuels Sources, Biofuel Policy, Biofuel Economy and Global Biofuel Projections," *Energy Conversion and Management* 49: 2106-16.
85. U.S. Energy Information Administration (2012) *Biofuels Issues and Trends* (Washington, DC: U.S. Department of Energy).
86. FAO (2009) *Global Agriculture Towards 2050*, www.fao.org/fileadmin/templates/wsfs/docs/Issues_papers/HLEF2050_Global_Agriculture.pdf, accessed August 26, 2014.
87. J. Clay (2012) "What Would it Take to Grow Enough Food While Reducing Environmental Impacts?" Interview in *Momentum, University of Minnesota* 4.1 (Winter 2012).
88. World Food Programme (2009) "World Must Double Food Production by 2050: FAO Chief," www.wfp.org/content/world-must-double-food-production-2050-fao-chief, accessed June 10, 2014.
89. FAO (2013) *Food Wastage Footprint. Impacts on Natural Resources*, www.fao.org/docrep/018/i3347e/i3347e.pdf, accessed August 26, 2014.
90. FAO (2013) *Food Wastage Footprint. Impacts on Natural Resources*, www.fao.org/docrep/018/i3347e/i3347e.pdf, accessed August 26, 2014.
91. J. Kruse (2010) "Estimating Demand for Agricultural Commodities to 2050," *Global Harvest Initiative*, http://globalharvestinitiative.org/Documents/Kruse%20-%20Demand%20for%20Agricultural%20Commoditites.pdf, accessed August 26, 2014.
92. FAO (2014) "FAO Food Price Index," www.fao.org/worldfoodsituation/foodprices index/en/, accessed May 9, 2014.
93. FAO (2014) *Annual Food Price Indices*, http://webcache.googleusercontent.com/search?q=cache:QyRhSF3LeogJ:www.fao.org/fileadmin/templates/worldfood/Reports_and_docs/Food_price_indices_data.xls+&cd=1&hl=en&ct=clnk&gl=nl, accessed August 26, 2014.
94. FAO (2012) *The State of Food Insecurity in the World 2012*, www.fao.org/docrep/016/i3027e/i3027e01.pdf, accessed August 26, 2014.
95. R. Zurayk (2011) "Use Your Loaf: Why Food Prices Were Crucial in the Arab Spring," *The Guardian*, July 17, 2011, www.theguardian.com/lifeandstyle/2011/jul/17/bread-food-arab-spring, accessed May 9, 2014.
96. M. Roig-Franzia (2007) "A Culinary and Cultural Staple in Crisis," *The Washington Post*, January 26, 2007, www.washingtonpost.com/wp-dyn/content/article/2007/01/26/AR2007012601896_pf.html, accessed May 9, 2014.
97. European Commission (2011) "Mexico–EU: Basic Statistical Indicators," http://epp.eurostat.ec.europa.eu/statistics_explained/index.php/Mexico-EU_-_basic_statistical_indicators, accessed May 9, 2014.
98. World Bank (2014) "World Development Indicators: Agricultural Inputs," http://wdi.worldbank.org/table/3.2, accessed May 9, 2014.

Chapter 3: Reading and understanding behavior in systems

99. The body of literature on systems dynamics and systems thinking is growing fast and developing over time. The founder of systems dynamics is Jay Forrester. Good applications of systems thinking are given in: P. Senge (1990) *The Fifth Discipline: The Art and Practice of the Learning Organization* (New York: Currency Doubleday); or, in Dutch, J. Schaveling, B. Bryan, and M. Goodman (2012) *Systeemdenken, van Goed Bedoeld Naar Goed Gedaan* (Academic Service).
100. D. Meadows (1999) *Leverage Points: Places to Intervene in a System* (The Sustainability Institute).

101. M. Ezekiel (1938) "The Cobweb Theorem," *The Quarterly Journal of Economics* 52.2: 255-80.
102. G. Hardin (1968) "The Tragedy of the Commons," *Science* 162.3859: 1243-48.
103. A. Grant (2013) *Give and Take: A Revolutionary Approach to Success* (London: Penguin).
104. P. Foster (2011) "Top 10 Chinese Food Scandals," *The Telegraph*, April 27, 2001, www.telegraph.co.uk/news/worldnews/asia/china/8476080/Top-10-Chinese-Food-Scandals.html, accessed May 7, 2014.

Chapter 4: Why do agricultural systems fail?

105. Holland Trade, "Holland is World-leading Exporter of Agri-food Products," www.hollandtrade.com/sector-information/agriculture-and-food/?bstnum=4909, accessed May 9, 2014.
106. Fairtrade (2013) "Powering Up Smallholder Farmers To Make Food Fair: A Five Point Agenda," www.fairtrade.net/fileadmin/user_upload/content/2009/news/2013-05-Fairtrade_Smallholder_Report_FairtradeInternational.pdf, accessed August 26, 2014.
107. Neumann Kaffee Gruppe, "About us," www.nkg.net/aboutus, accessed May 9, 2014.
108. G. Szala (2013) "The Smart Money Typically Are the Men Behind the Curtain. Who Are They and What Do They Do?" *Futures Magazine*, July 25, 2013, www.futuresmag.com/2013/07/25/10-top-global-commodity-trading-firms-smart-money, accessed May 9, 2014.
109. Cargill, "At a Glance," www.cargill.com/company/glance, accessed May 9, 2014.
110. Bunge, "Contact Us," www.bunge.com/contacts, accessed May 9, 2014.
111. Bunge (2012) *2012 Annual Report*, http://media.corporate-ir.net/media_files/irol/13/130024/AR2012/index.html, accessed August 26, 2014.
112. G. Szala (2013) "The Smart Money Typically Are the Men Behind the Curtain. Who Are They and What Do They Do?" *Futures Magazine*, July 25, 2013, www.futuresmag.com/2013/07/25/10-top-global-commodity-trading-firms-smart-money, accessed May 9, 2014.
113. S. Burnett (2013) "Gov. Quinn: No Tax Breaks for ADM Unless Pension Deal is Set," *Chicago Sun Times*, October 4, 2013, www.suntimes.com/22954864-761/gov-quinn-no-tax-breaks-for-adm-unless-pension-deal-is-set.html#.U2yM8PmSyE0, accessed May 9, 2014.
114. ADM, "ADM Facts," www.adm.com/en-US/company/Facts/Pages/default.aspx, accessed May 9, 2014.
115. G. Szala (2013) "The Smart Money Typically Are the Men Behind the Curtain. Who Are They and What Do They Do?" *Futures Magazine*, July 25, 2013, www.futuresmag.com/2013/07/25/10-top-global-commodity-trading-firms-smart-money, accessed May 9, 2014.
116. F. Lawrence (2011) "The Global Food Crisis: ABCD of Food—How the Multinationals Dominate Trade," *The Guardian*, June 2, 2011, www.theguardian.com/global-development/poverty-matters/2011/jun/02/abcd-food-giants-dominate-trade, accessed May 9, 2014.
117. Interview with Nick Goodall, Former CEO Bonsucro, August 6, 2013.
118. GRACE, "Animal Feed," www.sustainabletable.org/260/animal-feed, accessed June 3, 2014.
119. Interview with Hammad Naqi Khan, Global Cotton Leader WWF International, August 1, 2013.

120. FAO (1996) "Rome Declaration on World Food Security," *World Food Summit*, Rome, Italy, November 13–17, 1996.
121. J.C. Miller and K.H. Coble (2005) "Cheap Food Policy: Fact or Rhetoric?" paper prepared for the *American Agricultural Economics Association Annual Meeting*, Providence, Rhode Island, July 24–27, 2005.
122. *The Economist* (2007) "The End of Cheap Food," *The Economist*, December 6, 2007, www.economist.com/node/10252015, accessed July 8, 2014.
123. European Commission (2014) "The Early Years: Establishment of the CAP," http://ec.europa.eu/agriculture/cap-history/early-years/index_en.htm, accessed May 9, 2014.
124. M. Gallagher, M. Laver, and P. Mair (2006) *Representative Government in Modern Europe* (Singapore: McGraw-Hill Education [Asia], 4th edn): 142.
125. M. Curtis (2011) *Milking the Poor: How EU Subsidies Hurt Dairy Producers in Bangladesh* (Copenhagen: ActionAid).
126. R. Davison (2012) "The Five Worst Decisions Ever Made By the European Union," *The Conversation*, https://theconversation.com/the-five-worst-decisions-ever-made-by-the-european-union-9390, accessed May 9, 2014.
127. Trinity College Dublin (2010) "Exploring Links Between EU Agricultural Policy and World Poverty," www.tcd.ie/iiis/policycoherence/eu-agricultural-policy/protection-measures.php, accessed May 9, 2014.
128. K. Watkins and J. von Braun (2003) "Time to Stop Dumping on the World's Poor," Essay from IFPRI's 2002–2003 Annual Report," www.ifpri.org/sites/default/files/publications/ar02e1.pdf, accessed August 26, 2014.
129. M. Curtis (2011) *Milking the Poor: How EU Subsidies Hurt Dairy Producers in Bangladesh* (Copenhagen: ActionAid).
130. B. Waterfields (2009) "EU Butter Mountain to Return," *The Telegraph*, January 22, 2009, www.telegraph.co.uk/news/worldnews/europe/eu/4316726/EU-butter-mountain-to-return.html, accessed May 9, 2014.
131. ActionAid (2013) Farmgate: The Developmental Impact of Agricultural Subsidies, www.actionaid.org.uk/sites/default/files/doc_lib/farmgate_media_briefing.pdf, accessed August 26, 2014.
132. C. Hanrahan, B.A. Banks, and C. Canada (2011) *U.S. Agricultural Trade: Trends, Composition, Direction, and Policy* (Nova Science).
133. C. Hanrahan, B.A. Banks, and C. Canada (2011) *U.S. Agricultural Trade: Trends, Composition, Direction, and Policy* (Nova Science).
134. R. Johnson and J. Monke (2014) "What Is The Farm Bill?" *Congressional Research Service*, http://fas.org/sgp/crs/misc/RS22131.pdf, accessed August 26, 2014.
135. R.M. Chite (2013) *The 2013 Farm Bill: A Comparison of the Senate-Passed (S. 954) and House-Passed (H.R. 2642, H.R. 3102) Bills with Current Law*, http://fas.org/sgp/crs/misc/R43076.pdf, accessed August 26, 2014.
136. OECD (2001) "Towards More Liberal Agricultural Trade," *OECD Policy Brief*, www.oecd.org/tad/agricultural-trade/2674624.pdf, accessed August 26, 2014.
137. Interview with Frank Mechielsen, Policy Advisor, Oxfam Novib, June 12, 2013.
138. A. Ismi (2004) *Impoverishing a Continent: The World Bank and the IMF in Africa*, www.halifaxinitiative.org/updir/ImpoverishingAContinent.pdf, accessed August 26, 2014.
139. SAPRIN (2002) "The Policy Roots of Economic Crisis and Poverty: A Multi-Country Participatory Assessment of Structural Adjustment," www.saprin.org/SAPRIN_Synthesis_11-16-01.pdf, accessed August 26, 2014.

140. SAPRIN (2002) "The Policy Roots of Economic Crisis and Poverty: A Multi-Country Participatory Assessment of Structural Adjustment," www.saprin.org/SAPRIN_Synthesis_11-16-01.pdf, accessed August 26, 2014.
141. SAPRIN (2002) "The Policy Roots of Economic Crisis and Poverty: A Multi-Country Participatory Assessment of Structural Adjustment," www.saprin.org/SAPRIN_Synthesis_11-16-01.pdf, accessed August 26, 2014.
142. SAPRIN (2002) "The Policy Roots of Economic Crisis and Poverty: A Multi-Country Participatory Assessment of Structural Adjustment," www.saprin.org/SAPRIN_Synthesis_11-16-01.pdf, accessed August 26, 2014.
143. For instance: R. Hayman, T. Lawo, A. Crack, T. Kontinen, J. Okitoi, and B. Pratt (2013) *Legal Frameworks and Political Space for Non-Governmental Organisations: An Overview of Six Countries*, www.intrac.org/data/files/resources/771/Legal-Frameworks-and-Political-Space-for-Non-Governmental-Organisations-An-Overview-of-Six-Countries-July-2013.pdf, accessed August 26, 2014.
144. IAASTD (2009) "Agriculture at a Crossroads," www.fao.org/fileadmin/templates/est/Investment/Agriculture_at_a_Crossroads_Global_Report_IAASTD.pdf, accessed August 26, 2014.
145. Interview with Peter Erik Ywema, General Manager of the Sustainable Agricultural Initiative Platform, November 21, 2013.

Chapter 5: Phases of market transformation

146. A. Kahane (2004) *Solving Tough Problems: An Open Way of Talking, Listening and Creating New Realities* (San Francisco, CA: Berrett-Koehler Publishers, Inc.): 1. Kahane is a partner at Reos Partners, an international organization dedicated to supporting and building capacity for innovative collective action in complex social systems.
147. A. Kahane (2004) *Solving Tough Problems: An Open Way of Talking, Listening and Creating New Realities* (San Francisco, CA: Berrett-Koehler Publishers, Inc.): 2.
148. What happens on a smaller level also happens on a larger level. These three principles of successful and balanced systems are very similar to the three principles for successful relationships formulated by Bert Hellinger. Hellinger is a German psychotherapist associated with a therapeutic method best known as Family Constellations and Systemic Constellations. These three principles are called Connection, Order, and Balance and they define a balanced relationship within families and personal relationships.
149. Jujutsu means "Gentle Art," but can also be translated as "way of flexibility." It aims to defeat an attacker by manipulating or redirecting the attacker's force. Source: U.S. Ju-Jitsu Federation, www.usjjf.org/info/jujitsu.htm, accessed July 11, 2014.
150. J.D. Sterman (2002) "System Dynamics: Systems Thinking and Modeling for a Complex World," https://esd.mit.edu/WPS/internal-symposium/esd-wp-2003-01.13.pdf, accessed August 26, 2014.
151. J. Harich (2010) "Change Resistance as the Crux of the Environmental Sustainability Problem," *System Dynamics Review* 26.1 (January–March 2010): 35-72.
152. The term "level playing field" is used frequently and has various meanings depending on the context. Please note that, in this context, I use the term "level playing field" to indicate that all actors in a system effectively have to follow the same rules. I am not referring to the desired relationship between trader and producer.

Chapter 6: How does market transformation start?

153. L. Tacconi (2003) "Fires in Indonesia: Causes, Costs and Policy Implications," *CIFOR*, Occasional Paper No. 38.
154. R. Glastra, E. Wakker, and W. Richert (2002) *Oil Palm Plantations and Deforestation in Indonesia. What Role Do Europe and Germany Play?* (WWF Germany).
155. USAID is the United States federal government agency primarily responsible for administering civilian foreign aid.
156. E. Brown and M.F. Jacobson (2005) *Cruel Oil: How Palm Oil Harms Health, Rainforest & Wildlife* (Washington, DC: Center for Science in the Public Interest).
157. E. Wakker and J.W. van Gelder (2001) "Campaign Case Study: Dutch Banks and Indonesian Palm Oil," *Focus on Finance Newsletter*, www.profundo.nl/files/download/Focusonfinance0111.pdf, accessed August 27, 2014.
158. R. Glastra, E. Wakker, and W. Richert (2002) *Oil Palm Plantations and Deforestation in Indonesia: What Role Do Europe and Germany Play?* (WWF Germany).
159. N. Dudley, J.-P. Jeanrenaud, and S. Stolton (1997) *The Year the World Caught Fire*, www.equilibriumresearch.com/upload/document/theyeartheworldcaughtfire.pdf, accessed August 27, 2014.
160. E. Wakker (2000) *Funding Forest Destruction: The Involvement of Dutch Banks in the Financing of Oil Palm Plantations in Indonesia* (Greenpeace Netherlands).
161. Interview with Jan Maarten Dros, International Palm Oil Program Coordinator, Solidaridad, July 8, 2013.
162. E. Wakker (2000) *Funding Forest Destruction: The Involvement of Dutch Banks in the Financing of Oil Palm Plantations in Indonesia* (Greenpeace Netherlands).
163. E. Wakker and J.W. van Gelder (2001) "Campaign Case Study: Dutch Banks and Indonesian Palm Oil," *Focus on Finance Newsletter*, www.profundo.nl/files/download/Focusonfinance0111.pdf, accessed August 27, 2014.
164. Interview with Jan Maarten Dros, International Palm Oil Program Coordinator, Solidaridad, July 8, 2013.
165. Greenpeace, "Give the Orang-utan a Break," www.greenpeace.org.uk/files/po/index.html, accessed June 4, 2014.
166. Greenpeace, "Dirty Secret," http://dirtysecret.greenpeace.org, accessed June 4, 2014.
167. J. Hamprecht and D. Corsten (2006) *Supply Chain Strategy and Sustainability: The Migros Palm Oil Case*, http://oikos-international.org/wp-content/uploads/2013/10/oikos_Cases_2006_Migros.pdf, accessed August 27, 2014.
168. Migros did this successfully. The code was implemented and later extended to supplier. Eventually Migros succeeded in achieving a reduction of its palm oil purchasing volume by a third.
169. J. Hamprecht and D. Corsten (2006) "Supply Chain Strategy and Sustainability: The Migros Palm Oil Case," http://oikos-international.org/wp-content/uploads/2013/10/oikos_Cases_2006_Migros.pdf, accessed August 27, 2014.
170. Interview with Jan Kees Vis, Global Director of Sustainable Sourcing Development at Unilever and former Chair of the RSPO, September 12, 2013.
171. International Coffee Organization, "World Coffee Trade," www.ico.org/trade_e.asp, accessed June 4, 2014.
172. International Coffee Organization, "Mission," www.ico.org/mission.asp, accessed June 4, 2014.
173. Which Country?, "Which Country Drinks the Most Coffee?" www.whichcountry.co/which-country-drinks-the-most-coffee/, accessed June 4, 2014.

174. Smallholders own a small plot of land, typically smaller than two hectares. See also paragraph 2.1.
175. Fairtrade Foundation, "History," www.fairtrade.org.uk/what_is_fairtrade/history.aspx, accessed June 4, 2014.
176. Interview with Stefanie Miltenburg, Director International Corporate Social Responsibility at D.E Master Blenders 1753 and Director of the Douwe Egberts Foundation, June 21, 2013.
177. Interview with Annemieke Wijn, Senior Director for Commodity Sustainability Kraft Foods, August 5, 2013.
178. Interview with Nico Roozen, Managing Director Solidaridad, July 2, 2013.
179. Interview with Stefanie Miltenburg, Director International Corporate Social Responsibility at D.E Master Blenders 1753 and Director of the Douwe Egberts Foundation, June 21, 2013.
180. Interview with Stefanie Miltenburg, Director International Corporate Social Responsibility at D.E Master Blenders 1753 and Director of the Douwe Egberts Foundation, June 21, 2013.
181. Interview with Annemieke Wijn, Senior Director for Commodity Sustainability Kraft Foods, August 5, 2013.
182. B. Dirks (2003) "Douwe Egberts Koopt Bij Slavenplantages" ("Douwe Egberts Buys From Slave Plantations"), *De Volkskrant,* www.volkskrant.nl/vk/nl/2680/Economie/article/detail/756468/2003/04/14/Douwe-Egberts-koopt-bij-slavenplantages.dhtml, accessed June 6, 2014.
183. C. Gresser and S. Tickell (2004) *Mugged: Poverty in Your Coffee Cup* (Boston, MA: Oxfam America): 2.
184. Interview with Stefanie Miltenburg, Director International Corporate Social Responsibility at D.E Master Blenders 1753 and Director of the Douwe Egberts Foundation, June 21, 2013.
185. Interview with Annemieke Wijn, Senior Director for Commodity Sustainability Kraft Foods, August 5, 2013.
186. Make Chocolate Fair, "Cocoa Prices and Income of Farmers," http://makechocolate fair.org/issues/cocoa-prices-and-income-farmers-0, accessed June 11, 2014.
187. Interview with Bill Guyton, President WCF, July 12, 2013.
188. IITA, "Sustainable Tree Crops Program," www.iita.org/web/stcp/home;jsessionid=158 2248FB82E8C1D4B55E3D4531E9EA2, accessed June 16, 2014.
189. BBC (2000) "Cocoa Farm Slavery 'Exaggerated'," http://news.bbc.co.uk/2/hi/africa/948876.stm, accessed June 11, 2014.
190. S. Raghavan and S. Chatterjee (2001) "A Taste of Slavery: How Your Chocolate May be Tainted," www.rrojasdatabank.info/chocolate.pdf, accessed August 27, 2014.
191. D. Wolfe and Shazzie (2005) *Naked Chocolate* (Berkeley, CA: North Atlantic Books).
192. Payson Center (2010) *Fourth Annual Report. Oversight of Public and Private Initiatives to Eliminate the Worst Forms of Child Labor in the Cocoa Sector in Côte d'Ivoire and Ghana,* http://fairtrade.net/fileadmin/user_upload/content/2009/resources/2010-09-30_tulane-fourthann-cocoa-rprt.pdf, accessed August 27, 2014.
193. L. Kazemi (2011) *The Harkin-Engel Protocol: Ten Years On,* http://business-human rights.org/sites/default/files/media/documents/kazemi-re-harkin-engel-jun-2011.doc, accessed August 27, 2014.
194. This part is based on the following documents: Earth Rights International, "Amicus Brief in Doe v. Nestlé," www.earthrights.org/publication/amicus-brief-doe-v-Nestlé, accessed June 5, 2014; ILRF (2006) "On Halloween, Nestlé Claims no Responsibility for Child Labor," www.laborrights.org/stop-child-labor/cocoa-campaign/news/10993,

accessed June 5, 2014; L. Kazemi (2011) *The Harkin-Engel Protocol: Ten Years On*, http://business-humanrights.org/sites/default/files/media/documents/kazemi-re-harkin-engel-jun-2011.doc, accessed August 27, 2014; Payson Center (2010) *Fourth Annual Report. Oversight of Public and Private Initiatives to Eliminate the Worst Forms of Child Labor in the Cocoa Sector in Côte d'Ivoire and Ghana*, http://fairtrade.net/file admin/user_upload/content/2009/resources/2010-09-30_tulane-fourthann-cocoa-rprt .pdf, accessed August 27, 2014.

195. Interview with Daudi Lelijveld, Vice President of Barry Callebaut's Cocoa Sustainability Programs, August 8, 2013.
196. Smallstarter (2014) "Spices: How African Entrepreneurs Can Build A Business From This Old But Lucrative Product," www.smallstarter.com/browse-ideas/agribusiness-and-food/how-to-start-a-spice-business-in-africa, accessed July 14, 2014.

Chapter 7: The first mover and competition phase

197. Duurzaam-ondernemen.nl (2004) "Douwe Egberts Werkt Samen Met Utz Kapeh (Douwe Egberts Works Together With Utz Kapeh)," www.duurzaam-ondernemen.nl/douwe-egberts-werkt-samen-met-utz-kapeh/, accessed June 13, 2014.
198. Interview with Stefanie Miltenburg, Director International Corporate Social Responsibility at D.E Master Blenders 1753 and Director of the Douwe Egberts Foundation, June 21, 2013.
199. Interview with Annemieke Wijn, Senior Director for Commodity Sustainability Kraft Foods, August 5, 2013.
200. 4C Association, "History," www.4c-coffeeassociation.org/about-us/history.html#2, accessed June 4, 2014.
201. For more information about the founding of Max Havelaar see: N. Roozen and F. van der Hoff (2001) *Fairtrade: The Story behind Max Havelaar, Oké Bananas and Kuyichi Jeans (Fairtrade: Het Verhaal Achter Max Havelaar-koffie, Oké-bananen en Kuyichi-jeans)* (Amsterdam: Van Gennep).
202. Interview with Nico Roozen, Managing Director of Solidaridad, July 2, 2013.
203. The Dutch DOEN-Foundation, Hivos, and, later, Solidaridad deserve recognition for their leadership role in giving financial support to Utz Kapeh in those early days.
204. Interview with Carsten Schmitz-Hoffmann, Executive Director Private Sector Cooperation, GIZ, July 16, 2013.
205. RVO, "Van Oorsprong tot Kopje Koffie (Coffee: From Origin to Cup)," http://www.rvo .nl/node/8066, accessed June 4, 2014.
206. Organic data is not readily available. Numbers are based on: J. Pierrot, D. Giovannucci, and A. Kasterine (2011) *Trends in the Trade of Certified Coffees* (Geneva: International Trade Center); J. Potts, J. Meer, and J. Daitchman (2010) *The State of Sustainability Initiatives Review 2010: Sustainability and Transparency* (Winnipeg and London: IISD and IIED); S. Panhuysen and M. van Reenen (2012) *Coffee Barometer 2012* (The Hague: Tropical Commodity Coalition); UTZ Certified (2014) *UTZ Certified Impact Report January 2014* (The Netherlands: UTZ Certified Communications); 4C Association (2009) *Taking The Next Step. Annual Report 2009* (Germany: 4C Association); Fairtrade Labelling Organisations (FLO) (2012) *Monitoring the Scope and Benefits of Fairtrade* (Germany: Fairtrade International, 4th edn); 2007–2008 is the first coffee year of 4C.
207. Interview with Nico Roozen, Managing Director, Solidaridad, July 2, 2013.
208. Interview with David Rosenberg, Sustainability Officer at Ecom Agroindustrial Corp and former CEO of UTZ Certified, August 23, 2014.

209. S. Bowers (2009) "Sweet Deal: Dairy Milk to Carry Fairtrade Badge," *The Guardian*, March 4, 2009, www.theguardian.com/environment/2009/mar/04/cadbury-fair-trade-dairy-milk, accessed June 4, 2014.
210. Solidaridad, "Sustainable Cocoa," www.solidaridadnetwork.org/cocoa, accessed June 4, 2014.
211. Interview with Matthieu Guémas, Head of Business Development for Europe, Africa and Asia at GeoTraceability, June 17, 2013.
212. Mars, "Cocoa Certification," www.mars.com/gcc/en/principles-in-action/cocoa-certification.aspx, accessed June 16, 2014.
213. Mars, "Cocoa Certification," www.mars.com/gcc/en/principles-in-action/cocoa-certification.aspx, accessed June 16, 2014.
214. A. Lazo (2009) "Mars Sets Goal for Sustainable Cocoa Sources," *The Washington Post*, April 10, 2009, www.washingtonpost.com/wp-dyn/content/article/2009/04/09/AR2009040903943.html, accessed June 4, 2014.
215. CNN (2011) "The Dark Side of Chocolate," *The CNN Freedom Project*, http://thecnnfreedomproject.blogs.cnn.com/2011/04/06/the-dark-side-of-chocolate, accessed June 4, 2014.
216. Hershey, "BLISS Sustainable Cocoa," www.hersheys.com/bliss/our-story/rainforest-alliance.aspx, accessed June 4, 2014.
217. D. Ariosto (2012) "Hershey Pledges $10 Million to Improve West African Cocoa Farming, Fight Child Labor," *The CNN Freedom Project*, http://thecnnfreedomproject.blogs.cnn.com/2012/01/31/hershey-pledges-10-million-to-improve-west-african-cocoa-farming-fight-child-labor, accessed June 4, 2014.
218. Cargill, "The Cargill Cocoa Promise in Côte d'Ivoire," www.cargillcocoachocolate.com/sustainable-cocoa/our-promise-in-action/cote-d-ivoire/index.htm, accessed June 4, 2014.
219. TCC (2010) "Tropical Commodity Coalition Conference: Combining Results," www.teacoffeecocoa.org/tcc/content/download/420/2971/file/TCC_Combining%20Results_2010_lr.pdf, accessed August 27, 2014.
220. Interview with Nicko Debenham, Head of Cocoa at Armajaro, July 22, 2013.
221. Interview with Daudi Lelijveld, Vice President of Barry Callebaut's Cocoa Sustainability Programs, August 8, 2013).
222. M. Djama E. Fouilleux, and I. Vagneron (2011) "Standard-setting, Certifying and Benchmarking: A Governmentality Approach to Sustainability in the Agro-Food Sector," in S. Ponte, P. Gibbon, and J. Vestergaard (eds.), *Governing Through Standards* (London: Palgrave Macmillan): 184-209.
223. R. Glastra, E. Wakker, and W. Richert (2002) *Oil Palm Plantations and Deforestation in Indonesia. What Role Do Europe and Germany Play?* (WWF Germany).
224. RSPO, "History," www.rspo.org/en/history, accessed July 3, 2014.
225. Interview with Darrel Webber, RSPO Secretary General, September 19, 2013.
226. Interview with Darrel Webber, RSPO Secretary General, September 19, 2013.
227. Interview with Darrel Webber, RSPO Secretary General, September 19, 2013.
228. Interview with Jan Maarten Dros, International Palm Oil Program Coordinator, Solidaridad, July 8, 2013.
229. M. Djama E. Fouilleux, and I. Vagneron (2011) "Standard-setting, Certifying and Benchmarking: A Governmentality Approach to Sustainability in the Agro-Food Sector," in S. Ponte, P. Gibbon, and J. Vestergaard (eds.), *Governing Through Standards* (London: Palgrave Macmillan): 184-209.

230. D. Sheil, A. Casson, E. Meijaard, M. van Noordwjik, J. Gaskell, J. Sunderland-Groves, K. Wertz, and M. Kanninen (2009) "The Impacts and Opportunities of Oil Palm in Southeast Asia: What Do We Know and What Do We Need to Know?" *CIFOR*, Occasional paper no. 51.
231. D.H. Goenadi (2008) "Perspective on Indonesian Palm Oil Production," paper presented at the *International Food & Agricultural Trade Policy Council's Spring 2008 Meeting*, Bogor, Indonesia, May 12, 2008.
232. Interview with Darrel Webber, RSPO Secretary General, September 19, 2013.
233. Friends of the Earth (2005), *The Oil for Ape Scandal: How Palm Oil Threatens Orangutan Survival* (London: Friends of the Earth Trust).
234. Friends of the Earth (2008) "Palm Oil Campaign Wins Award," www.foe.co.uk/news/palm_oil, accessed June 5, 2014.
235. J.W. Van Gelder and E. Wakker (2006) *People, Planet, Palm Oil? A Review of the Oil Palm and Forest Policies Adopted by Dutch Banks* (Amsterdam: Milieudefensie—Friends of the Earth Netherlands).
236. BankTrack (2005) "Unproven Equator Principles. A BankTrack Statement," www.banktrack.org/download/unproven_equator_principles_a_banktrack_statement, accessed August 27, 2014.
237. M. Naguran (2013) *The Haze, Hazier Deals and the Fiery Blame Game*, www.gaiadiscovery.com/agriculture-industry/the-haze-hazier-deals-and-the-fiery-blame-game.html, accessed August 27, 2014.
238. Greenpeace (2008) "The Climate Bomb is Ticking: Call for Zero Deforestation to Protect the Climate," Policy Statement: 4.
239. *The Economist* (2010) "The Other Oil Spill," www.economist.com/node/16423833, accessed August 27, 2014.
240. Greenpeace (2011) "Giant Indonesian Palm Oil Company Announces Plan to Halt Forest Destruction," www.greenpeace.org/seasia/news/Palm-Oil-Company-Announces-Plan-to-Halt-Forest-Destruction, accessed June 5, 2014.
241. *Environmental Leader* (2010) "Nestlé Quits Sinar Mas After Greenpeace Campaign," www.environmentalleader.com/2010/05/18/Nestlé-quits-sinar-mas-after-greenpeace-campaign/, accessed June 5, 2014.
242. Y. Basiron (2007) "Palm Oil Production Through Sustainable Plantations," *European Journal of Lipid Science and Technology* 109.4: 289-295.
243. *The Jakarta Post* (2010) "Indonesia Develops Rival Sustainable Palm Oil Scheme," *The Jakarta Post*, November 10, 2010, www.thejakartapost.com/news/2010/11/10/indonesia-develops-rival-sustainable-palm-oil-scheme.html, accessed June 5, 2014.
244. L. Morrissey (2011) "Seeing REDD in Indonesia," *Business Spectator*, June 2, 2011, www.businessspectator.com.au/article/2011/6/1/policy-politics/seeing-redd-indonesia, accessed August 27, 2014.
245. Interview with Jan Maarten Dros, International Palm Oil Program Coordinator, Solidaridad, July 8, 2013.
246. Interview with Jan Kees Vis, Global Director of Sustainable Sourcing Development Unilever and former chair of the RSPO, September 12, 2013.
247. ISCC, "ISCC Offers Solutions for Different Markets," www.iscc-system.org/en/iscc-system/about-iscc, accessed June 5, 2014.
248. D. Sheil, A. Casson, E. Meijaard, M. Van Noordwijk, J. Gaskell, J. Sunderland-Groves, and M. Kanninen (2009) "The Impacts and Opportunities of Oil Palm in Southeast Asia: What Do We Know and What Do We Need to Know?" *CIFOR*, Occasional paper no. 51.

249. Taskforce Sustainable Palm Oil (2010) *Manifesto of the Task Force Sustainable Palm Oil* (Rijswijk: Task Force Duurzame Palm Olie).
250. Taskforce Sustainable Palm Oil (2010) "Alle voor de NL-Markt Bestemde Palmolie is Duurzaam in 2015 (All the Palm Oil Destined for the Dutch Market Bill be Sustainable in 2015)," press release, www.taskforceduurzamepalmolie.nl/Portals/4/download/Persbericht_TFDPO-02112010.pdf, accessed June 5, 2015.
251. SustainableBusiness.com (2013) "Starbucks Agrees to 100% Certified Palm Oil," www.sustainablebusiness.com/index.cfm/go/news.display/id/24602, accessed June 5, 2014.
252. Entelliprise (2013) "Blommer to Offer 100% RSPO-certified Palm Oil in U.S.," www.entelliprise.com/External/News_full.aspx?DocNumber=6694, accessed June 5, 2014.
253. Innofood (2013) "100% Duurzame en Traceerbare Palmolie in Nutella (100% Sustainable and Traceable Palm Oil in Nutella)," www.innofood.org/nl/nieuws/11009/100-duurzame-en-traceerbare-palmolie-in-nutella.html, accessed June 5, 2014.
254. Interview with Bruce Wise, Global Product Specialist, Sustainable Business Advisory, International Finance Corporation, July 11, 2013.
255. RSPO, "Who is RSPO?" www.rspo.org/en/who_is_rspo, accessed June 5, 2014.
256. N. Sizer, J. Anderson, F. Stolle, S. Minnemeyer, M. Higgins, A. Leach, and A. Alisjahbana (2014) "Fires in Indonesia at Highest Levels Since 2013 Haze Emergency," *The Guardian*, March 14, 2014, www.theguardian.com/environment/2014/mar/14/fires-indonesia-highest-levels-2012-haze-emergency, accessed June 5, 2014.
257. H. Davidson (2013) "RSPO Members Implicated in Air Pollution Crisis, Says Greenpeace," *The Guardian*, July 12, 2013, www.theguardian.com/environment/2013/jul/12/sustainable-palm-oil-pollution-crisis, accessed June 5, 2014.
258. A. Gangopadhay and B. Otto (2013) "Indonesia Plans to Ratify Haze Pact," *The Wall Street Journal*, July 17, 2013, http://online.wsj.com/news/articles/SB10001424127887324448104578611241859487394, accessed June 5, 2014.
259. Interview with Jan Kees Vis, Global Director of Sustainable Sourcing Development at Unilever and former chair of the RSPO, September 12, 2013.
260. Tropical Commodity Coalition (2010) "Tea Barometer," www.teacoffeecocoa.org/tcc/content/download/404/2879/file/TCC_TEA_baro2010_LP.pdf, accessed August 27, 2014.
261. FAO, "FAOstat," http://faostat.fao.org, accessed August 27, 2014.
262. Rainforest Alliance, "Our Work with Unilever," http://www.rainforest-alliance.org/about/company-commitments/unilever, accessed August 31, 2014.
263. M. Grossman (2011) Tea Sector Overview, www.idhsustainabletrade.com/thee-tea-trade-flow, accessed September 1, 2014.
264. FAO, "FAOstat," http://faostat.fao.org, accessed August 27, 2014.
265. Based on: Animal Welfare Approved, "About," http://animalwelfareapproved.org/about, accessed July 9, 2014; C. Brooks "Consequences of Increased Global Meat Consumption on the Global Environment: Trade in Virtual Water, Energy and Nutrients," *Stanford Woods Institute for the Environment*, https://woods.stanford.edu/environmental-venture-projects/consequences-increased-global-meat-consumption-global-environment, accessed July 9, 2014; FAO (2006) "Livestock Impacts on the Environment," www.fao.org/ag/magazine/0612sp1.htm, accessed July 9, 2014; Standards Map, *ITC*, www.standardsmap.org/identify, accessed July 9, 2014; SAI Platform, "SAI Platform Launches New Principles for Sustainable Beef Farming," www.saiplatform.org/pressroom/101/33/SAI-Platform-launches-new-Principles-for-Sustainable-Beef-Farming, accessed July 9, 2014.
266. FAO, "FAOstat," http://faostat.fao.org, accessed August 27, 2014.

267. KPMG (2013) "Sustainable Insight: A Roadmap to Responsible Soy," www.kpmg.com/Global/en/IssuesAndInsights/ArticlesPublications/sustainable-insight/Documents/roadmap-responsible-soy-v2.pdf, accessed August 27, 2014; Fairfood, "Soy," http://test.fairfood.org/research/production-chains/soy, accessed July 8, 2014.
268. *African Business Magazine* (2012) "The Global Flower Trade," http://africanbusinessmagazine.com/sector-reports/agriculture/the-global-flower-trade, accessed July 11, 2014.
269. Direct Floral Source, "Israeli Flower Market," www.directfloralintl.com/israel.html, accessed July 11, 2014.
270. B. Gollnow (2002) "Exporting Cut Flowers," *New South Wales Government*, www.dpi.nsw.gov.au/agriculture/horticulture/floriculture/industry/export, accessed July 11, 2014.
271. FAO (2012) "The State of World Fisheries and Aquaculture," www.fao.org/docrep/016/i2727e/i2727e.pdf, accessed August 27, 2014.
272. Greenpeace, "Sustainable Aquaculture," www.greenpeace.org/international/en/campaigns/oceans/sustainable-aquaculture, accessed July 8, 2014; M. Allsopp, P. Johnston, and D. Santillo (2008) Challenging the Aquaculture Industry on Sustainability. Technical Overview (Amsterdam: Greenpeace).
273. Information in this paragraph is based on: Sustainable Sugarcane Initiative (SSI), http://sri-india.net/ssi_website/index.html, accessed July 8, 2014; WWF, "Better Sugarcane Initiative (BSI)," http://wwf.panda.org/what_we_do/how_we_work/businesses/transforming_markets/solutions/bettermarkets/farming/sugarcane2/bonsucro/better_sugarcane_initiative, accessed August 27, 2014; C.V. Xavier, F.T. Pitta, and M.L. Mendonça (2011) "A Monopoly in Ethanol Production in Brazil: The Cosan–Shell Merger," TNI, www.social.org.br/ethanol_monopoly_brazil.pdf, accessed August 27, 2014; WWF, "Sugarcane," www.worldwildlife.org/industries/sugarcane, accessed July 8, 2014; WWF (2012) "The 2050 Criteria Guide to Responsible Investment in Agricultural, Forest, and Seafood Commodities," http://awsassets.panda.org/downloads/the_2050_critera_report.pdf, accessed August 27, 2014; N. Gray (2013) "Unilever Joins Solidaridad to Tackle Sustainable Sugar Cane Challenge," www.foodnavigator.com/Business/Unilever-joins-with-Solidaridad-to-tackle-sustainable-sugar-cane-challenge, accessed July 8, 2014; Global Voices (2013) "New Sugar Cane Farming Bill in Brazil Threatens Amazon," http://globalvoicesonline.org/2013/07/12/new-sugar-cane-farming-bill-in-brazil-threatens-amazon, accessed July 8, 2014; ELLA, "Sugarcane Agro-ecological Zoning: Greening the Expansion of Ethanol," *Policy Brief*, http://ella.practicalaction.org/node/1125#, accessed August 27, 2014.
274. USDA (2014) "Cotton: World Markets and Trade. April 2014," www.thecropsite.com/reports/?id=3674, accessed August 27, 2014.
275. IDH, "Better Cotton Fast Track Program," www.idhsustainabletrade.com/site/getfile.php?id=365, accessed August 27, 2014.
276. T.P. Townsend (2010) *Report of the Executive Director to the 69th Plenary Meeting of the International Cotton Advisory Committee* (Lubbock: ICAC).
277. Information in this paragraph is based on: Y. Montouroy (2013) "The EU FLEGT Action Plan to Counter Illegal Logging: Recentralization of European Rule Making, International Cooperation and Privatized Global Forest Governance," *ICCP*, www.icpublicpolicy.org/IMG/pdf/panel_46_s3_yves_montouroy.pdf, accessed August 27, 2014; FAO (2012) "State of the World's Forests," www.fao.org/docrep/016/i3010e/i3010e.pdf, accessed August 27, 2014; IDH, "Mainstreaming Sustainability in Tropical Timber," *Position Paper*, www.idhsustainabletrade.com/site/getfile.php?id=338, accessed

August 27, 2014; M. van Oorschot, M. Kok, J. Brons, S. van der Esch, J. Janse, T. Rood, E. Vixseboxse, H. Wilting, and W. Vermeulen (2013) "Verduurzaming van International Handelsketens (Making International Supply Chains More Sustainable)," *PBL*, www.pbl.nl/sites/default/files/cms/publicaties/PBL_2013_Verduurzaming%20van%20handelsketens_630.pdf, accessed August 27, 2014.

278. WWF, "Illegal Logging," wwf.panda.org/about_our_earth/about_forests/deforestation/forest_illegal_logging, accessed July 11, 2014.
279. Interview with Stefanie Miltenburg, Director International Corporate Social Responsibility at D.E Master Blenders 1753 and Director of the Douwe Egberts Foundation, June 21, 2013.
280. Interview with Rob Cameron, former CEO Fairtrade International, October 8, 2013.
281. Interview with Han de Groot, Executive Director of UTZ Certified, June 25, 2013.
282. Interview with Peter Erik Ywema, General Manager of the Sustainable Agricultural Initiative Platform, November 21, 2013.
283. H. Fuchs (2014) "Sumatra's Burning Rainforests," *DW*, www.dw.de/sumatras-burning-rainforests/g-17489217, accessed June 5, 2014.

Chapter 8: The critical mass and institutionalization phase

284. Information based on the following: Barry Callebaut launched Cocoa Horizons in March 2012, a CHF40 million cocoa sustainability initiative to boost farm productivity, increase quality, and improve family livelihoods in key cocoa producing countries over ten years. See: www.barry-callebaut.com/cocoa-horizons, accessed July 8, 2014; Mondelēz International announced its Cocoa Life program in 2012. With investment of $400 million, it aims to improve the livelihoods of cocoa farmers. See: O. Nieburg (2012) "Mondelēz Pumps $400m in Sustainable Cocoa supply chain," www.confectionerynews.com/Commodities/Mondelez-pumps-400m-in-sustainable-cocoa-supply-chain, accessed July 8, 2014; CocoaLink is a program started by Hershey in 2011 to help farmers improve production and quality of cocoa through farm training provided by low-cost SMS messages and expert exchanges via mobile phones. See: *Business Wire* (2012) "Hershey's CocoaLink Mobile Phone Program Delivers 100,000 Farmer and Family Messages During First Year in Ghana," *Business Wire*, August 6, 2012, www.businesswire.com/news/home/20120806005736/en/Hershey%E2%80%99s-CocoaLink-Mobile-Phone-Program-Delivers-100000#.U7wEg_mSyE0, accessed July 8, 2014; WCF Cocoa Livelihoods Program is a five-year, $40 million initiative that began in 2009 to increase farmers' incomes by enhancing their knowledge of agricultural techniques. The program is funded by the Bill and Melinda Gates Foundation and 15 chocolate sector companies. See: GCNF (2012) "The Hershey Company," www.gcnf.org/spotlight/the-hershey-company, accessed July 8, 2014; Mars committed $30 million in 2011 to its Sustainable Cocoa Initiative which focuses on research, certification, and direct intervention programs to boost productivity. See: Multivu (2012) "Mars Chocolate Invests in Creating a Sustainable Cocoa Industry," www.multivu.com/mnr/58400-mars-chocolate-invests-in-creating-a-sustainable-cocoa-industry, accessed July 8, 2014; Processors Alliance for Cocoa Traceability and Sustainability (PACTS), a $3 million joint venture between Blommer, Cemoi Chocolatiers, and Petra Foods, started in 2011 to improve the supply of high-quality, fermented cocoa beans from the Ivory Coast while improving the livelihoods of the local cocoa farming community. See: Blommer (2010) "Blommer Chocolate Announces Cocoa Sustainability Partnership Supporting Ivory Coast Farmers," www.blommer.com/_documents/Blommer-CocoaAction-PR_6-9-14.pdf,

accessed July 8, 2014; The Cargill Cocoa Promise, launched in 2012, builds on Cargill's commitment to increase farmer yields and income through farmer training, community support, and farm development. See: Peter's Chocolate (2012) "The Cargill Cocoa Promise Strengthens Company's Commitment to Sustainable Cocoa and its Support for Cocoa Farmers and Communities," www.cargill.com/news/releases/2012/NA3069037.jsp, accessed July 8, 2014; Cocoa Productivity and Quality Program (CPQP) was launched by IDH to mainstream innovations on effective farmer support and improved production. See: IDH the sustainable trade initiative, "CPQP," www.idhsustainabletrade.com/CPQP, accessed July 8, 2014.

285. I. Almeida (2013) "Goldman Sachs Sees 2012–13 Global Cocoa Shortage of 100,000 Tons," www.bloomberg.com/news/2013-01-14/goldman-sachs-sees-2012-13-global-cocoa-shortage-of-100-000-tons.html, accessed July 8, 2014.
286. Interview during a conference with a respondent who wished to remain anonymous, January 2014.
287. Interview with Bill Guyton, President WCF, July 12, 2013.
288. Interview with Bill Guyton, President WCF, July 12, 2013.
289. WCF (2014) "Global Chocolate and Cocoa Companies Announce Unprecedented Sustainability Strategy in Côte d'Ivoire," http://worldcocoafoundation.org/global-chocolate-and-cocoa-companies-announce-unprecedented-sustainability-strategy-in-cote-divoire, accessed August 27, 2014.
290. Interview with Leif Pedersen, Senior Commodities Advisor UNDP, August 2, 2013.
291. International Coffee Organization (2014) "World Coffee Trade (1963–2013): A Review of the Markets, Challenges and Opportunities Facing the Sector," www.ico.org/news/icc-111-5-r1e-world-coffee-outlook.pdf, accessed August 27, 2014.
292. COSA (2013) *The COSA Measuring Sustainability Report: Coffee and Cocoa in 12 Countries* (Philadelphia: The Committee on Sustainability Assessment).
293. S. Panhuysen and M. van Reenen (2012) *Coffee Barometer 2012* (The Hague: Tropical Commodity Coalition).
294. M. Groeg-Gran (2005) *From Bean to Cup: How Consumer Choice Impacts on Coffee Producers and the Environment* (London: Consumers International).
295. K. Laroche and B. Guittard (2009) *The Impact of Fairtrade Labelling on Small-scale Producers. Conclusions of the First Studies*, www.fairtrade.at/fileadmin/user_upload/PDFs/Produzenten/2009_MHF_The_impact_of_Fairtrade_labelling_on_small-scale_producers.pdf, accessed August 27, 2014.
296. Steering Committee of the State-of-Knowledge Assessment of Standards and Certification (2012) *Toward Sustainability: The Roles and Limitations of Certification* (Washington, DC: RESOLVE).
297. COSA (2013) *The COSA Measuring Sustainability Report: Coffee and Cocoa in 12 Countries* (Philadelphia: The Committee on Sustainability Assessment).
298. J. Potts, M. Lynch, A. Wilkings, G.A. Huppé, M. Cunningham, and V. Voora (2014) *The State of Sustainability Initiatives Review 2014: Standards and the Green Economy* (Winnipeg and London: IISD and IIED).
299. Interview with Stefanie Miltenburg, Director International Corporate Social Responsibility at D.E Master Blenders 1753 and Director of the Douwe Egberts Foundation, June 21, 2013.
300. IDH, the sustainable trade initiative, "What We Do," www.idhsustainabletrade.com/what-we-do, accessed July 8, 2014.
301. Interview with Joost Oorthuizen, Executive Director IDH, November 12, 2013.
302. Interview with Joost Oorthuizen, Executive Director IDH, November 12, 2013.

303. Interview with Stefanie Miltenburg, Director International Corporate Social Responsibility at D.E Master Blenders 1753 and Director of the Douwe Egberts Foundation, June 21, 2013.
304. Interview with Cornel Kuhrt, coffee industry expert, September 9, 2013.
305. Interview with Carsten Schmitz-Hoffmann, former Senior Executive, 4C, July 16, 2013.
306. Interview with Peter Erik Ywema, General Manager of the Sustainable Agricultural Initiative Platform, November 21, 2013.
307. Interview with Lise Melvin, former CEO, Better Cotton Initiative, August 19, 2013.
308. Interview with Annemieke Wijn, former Senior Director for Commodity Sustainability, Kraft Foods, August 5, 2013.
309. Oxfam America, "Behind the Brands," www.behindthebrands.org, accessed July 8, 2014.

Chapter 9: The level playing field phase

310. European Commission (2011) "A Common Mobile Phone Charger: Questions and Answers," http://europa.eu/rapid/press-release_MEMO-11-75_en.htm?locale=en, accessed July 8, 2014.
311. G. Peev (2014) "All Mobile Phones in EU to Have the Same Charger: European Parliament Votes on Law in Attempt to Cut Down Electronic Clutter," *Daily Mail*, March 14, 2014, www.dailymail.co.uk/news/article-2580597/All-mobile-phones-EU-charger-European-Parliament-votes-law-attempt-cut-electronic-clutter.html, accessed July 8, 2014.
312. Information in this paragraph is based on: European Commission (2011) "CE Marking Makes Europe's Market Yours!" http://ec.europa.eu/enterprise/policies/single-market-goods/cemarking/downloads/ce_leaflet_economic_operators_en.pdf, accessed August 27, 2014; European Commission, "Energy," http://ec.europa.eu/energy/lumen/faq/index_en.htm, accessed July 8, 2014; L.R. Brown (2008) "Chapter 11: Raising Energy Efficiency: Banning the Bulb," www.earth-policy.org/books/pb3/PB3ch11_ss2, accessed August 27, 2014, in L.R. Brown, *Plan B 3.0: Mobilizing to Save Civilization* (New York: W.W. Norton & Company).
313. Greenwashing Lamps (2009) "The Global Anti-Lightbulb Campaign," http://greenwashinglamps.wordpress.com/2009/03/29/anti-lightbulb-campaign, accessed July 8, 2014.
314. Energy Star, "Product Retrospective: Residential Lighting," *U.S. Environmental Protection Agency*, www.energystar.gov/ia/products/downloads/Residential_Lighting_Highlights.pdf, accessed July 8, 2014.
315. Energy Saving Trust (2013) "Energy Saving Trust to Extend Labelling and Verification Schemes," www.energysavingtrust.org.uk/Energy-Saving-Trust/Press/Press-releases/Energy-Saving-Trust-to-extend-labelling-and-verification-schemes, accessed July 8, 2014.
316. L.R. Brown (2008) "Chapter 11: Raising Energy Efficiency: Banning the Bulb," www.earth-policy.org/books/pb3/PB3ch11_ss2, accessed August 27, 2014, in L.R. Brown, *Plan B 3.0: Mobilizing to Save Civilization* (New York: W.W. Norton & Company).
317. L.R. Brown (2008) "Chapter 11: Raising Energy Efficiency: Banning the Bulb," www.earth-policy.org/books/pb3/PB3ch11_ss2, accessed August 27, 2014, in L.R. Brown, *Plan B 3.0: Mobilizing to Save Civilization* (New York: W.W. Norton & Company).
318. Unique Lights, "EU Law & Energy Market," www.uniquelights.net/index.php?option=com_content&view=article&id=32&Itemid=29&lang=en, accessed July 8, 2014.

319. Fox News (2013) "Lights Out for the Incandescent Light Bulb as of Jan. 1, 2014," Fox News, December 31, 2013, www.foxnews.com/tech/2013/12/31/end-road-for-incandescent-light-bulb, accessed July 8, 2014.
320. European Commission, "Energy," http://ec.europa.eu/energy/lumen/faq/index_en.htm, accessed July 8, 2014.
321. China.org (2012) "China Moves to Phase Out Incandescent Bulbs," China.org, October 18, 2012, www.china.org.cn/environment/2012-10/18/content_26829256.htm, accessed July 8, 2014.

Chapter 10: Twelve questions about the market transformation curve

322. Unilever, "Cage-Free Eggs and Sustainable Dairy," www.unilever.com/sustainable-living-2014/reducing-environmental-impact/sustainable-sourcing/cage-free-eggs-and-sustainable-dairy, accessed July 8, 2014.
323. A good article defining and describing the value of social entrepreneurship as a source of innovation and value creation is: F. Santos (2012) "A Positive Theory of Social Entrepreneurship," *Journal of Business Ethics* 111.3: 335-51.
324. Everyone a change-maker is the motto of the Ashoka network-innovators for the public. See: www.ashoka.org/visionmission.
325. Schwab Foundation for Social Entrepreneurship, www.schwabfound.org, accessed August 28, 2014.
326. Echoing Green, www.echoinggreen.org, accessed August 28, 2014.
327. Unreasonable Institute, http://unreasonableinstitute.org, accessed August 28, 2014.
328. UnLtd, https://unltd.org.uk, accessed August 28, 2014.

Ten examples of change-makers

329. GM Watch (2013) "Pesticide Illness Triggers Anti-Monsanto Protest in Argentina," www.gmwatch.org/index.php/news/archive/2013/15134-pesticide-illness-triggers-anti-monsanto-protest-in-argentina, accessed July 8, 2014.
330. GM Watch (2013) "Pesticide Illness Triggers Anti-Monsanto Protest in Argentina," www.gmwatch.org/index.php/news/archive/2013/15134-pesticide-illness-triggers-anti-monsanto-protest-in-argentina, accessed July 8, 2014.
331. L. Graves (2012) "Sofia Gatica, Argentine Activist, Faced Anonymous Death Threats For Fighting Monsanto Herbicide," *Huffington Post*, March 5, 2012, www.huffington post.com/2012/05/03/argentine-activist-sofia-gatica-monsanto_n_1475659.html, accessed July 8, 2014.
332. D. Datta (2013) "The Bribe Republic," *India Today*, July 12, 2013, http://indiatoday.intoday.in/story/bribery-in-india-bribe-republic-anti-corruption-bureau/1/291074.html, accessed July 8, 2014.
333. N. Singh (2010) "The Trillion-dollar Question," *The Financial Express*, December 19, 2010, www.financialexpress.com/news/the-trilliondollar-question/726482/0, accessed July 8, 2014.
334. J. Bajoria (2011) "Corruption Threatens India's Growth," *Council on Foreign Relations*, www.cfr.org/india/corruption-threatens-indias-growth/p24259, accessed July 8, 2014.
335. M. Chêne (2009) "Overview of Corruption and Anti-Corruption Efforts in India," *U4*, www.u4.no/publications/overview-of-corruption-and-anti-corruption-efforts-in-india, accessed July 8, 2014.

336. Bloomberg View (2011) "Indians Divide Over Policing a Watchdog: World View," www.bloombergview.com/articles/2011-06-21/indians-divide-over-policing-a-watchdog-world-view, accessed July 8, 2014.
337. A. Kumar (2011) "The Story of the 'India Against Corruption' Campaign," www.socialism.in/index.php/the-story-of-the-india-against-corruption-campaign, accessed July 8, 2014.
338. A. Kumar (2011) "The Story of the 'India Against Corruption' Campaign," www.socialism.in/index.php/the-story-of-the-india-against-corruption-campaign, accessed July 8, 2014.
339. KPMG (2011) "Survey on Bribery and Corruption. Impact on Economy and Business Environment," www.kpmg.com/Global/en/IssuesAndInsights/ArticlesPublications/Documents/bribery-corruption.pdf, accessed August 27, 2014.
340. Ashoka, "Bambang Suwerda," www.ashoka.org/fellow/bambang-suwerda, accessed July 8, 2014.
341. E. Munawar and J. Fellner, "Injury Time for Indonesian Landfills," *Waste Management World* 14.2, www.waste-management-world.com/articles/print/volume-14/issue-2/features/injury-time-for-indonesian-landfills.html, accessed July 8, 2104.
342. Ashoka, "Bambang Suwerda," www.ashoka.org/fellow/bambang-suwerda, accessed July 8, 2014.
343. Ashoka, "Bambang Suwerda," www.ashoka.org/fellow/bambang-suwerda, accessed July 8, 2014.
344. Ashoka, "Bambang Suwerda," www.ashoka.org/fellow/bambang-suwerda, accessed July 8, 2014.
345. Ashoka, "Bambang Suwerda," www.ashoka.org/fellow/bambang-suwerda, accessed July 8, 2014.
346. *Daily Sabah* (2014) "743 Million Tons of Soil Perishes in Turkey Due to Erosion," *Daily Sabah*, June 19, 2014, www.dailysabah.com/science/2014/06/19/743-million-tons-of-soil-perishes-in-turkey-due-to-erosion, accessed July 8, 2014.
347. Ashoka, "Hayrettin Karaca," www.ashoka.org/fellow/hayrettin-karaca, accessed July 8, 2014.
348. Ashoka, "Hayrettin Karaca," www.ashoka.org/fellow/hayrettin-karaca, accessed July 8, 2014.
349. A.K. Raymond (2011) "A Brief History of Drunk Driving," *The Fix*, November 23, 2011, www.thefix.com/content/brief-history-drunk-driving-dui-laws-thanksgiving7007, accessed July 8, 2014.
350. D.J. Hanson, "Candy Lightner," www2.potsdam.edu/alcohol/Controversies/1119636699.html#.U6rEYfmSyE2, accessed July 8, 2014.
351. D.J. Hanson, "Candy Lightner," www2.potsdam.edu/alcohol/Controversies/1119636699.html#.U6rEYfmSyE2, accessed July 8, 2014.
352. A.K. Raymond (2011) "A Brief History of Drunk Driving," *The Fix*, November 23, 2011, www.thefix.com/content/brief-history-drunk-driving-dui-laws-thanksgiving7007, accessed July 8, 2014.
353. A.K. Raymond (2011) "A Brief History of Drunk Driving," *The Fix*, November 23, 2011, www.thefix.com/content/brief-history-drunk-driving-dui-laws-thanksgiving7007, accessed July 8, 2014.
354. D. Hohn (2008) "Sea of Trash," *The New York Times*, June 22, 2008, www.nytimes.com/2008/06/22/magazine/22Plastics-t.html?pagewanted=3&sq&st=cse%22&scp=4%22donovan%20hohn&_r=0, accessed July 8, 2014.
355. Algalita, www.algalita.org, accessed August 28, 2014.

356. C. Summers (2012) "What Should Be Done About Plastic Bags?" *BBC*, www.bbc.com/news/magazine-17027990, accessed July 8, 2014.
357. European Commission (2013) "Environment: Commission Proposes to Reduce the Use of Plastic Bags," http://europa.eu/rapid/press-release_IP-13-1017_en.htm, accessed July 8, 2013.
358. Healthcare Packaging (2012) "Fighting Counterfeit Pharmaceuticals: A Primer," www.healthcarepackaging.com/trends-and-issues/traceability-and-authentication/fighting-counterfeit-pharmaceuticals-primer, accessed July 8, 2014.
359. D. Talbot (2013) "The mPedigree Network, Based in Ghana, Lets People Determine With a Text Message Whether Their Medicine is Legitimate," *Technology Review*, www.technologyreview.com/lists/innovators-under-35/2013/entrepreneur/bright-simons, accessed July 8, 2014.
360. D. Talbot (2013) "The mPedigree Network, Based in Ghana, Lets People Determine With a Text Message Whether Their Medicine is Legitimate," *Technology Review*, www.technologyreview.com/lists/innovators-under-35/2013/entrepreneur/bright-simons, accessed July 8, 2014.
361. mPedigree, "Where We Are & Where We Want to Go," http://mpedigree.net/mpedigreenet/index.php/where-we-are, accessed July 8, 2014.
362. D. Talbot (2013) "The mPedigree Network, Based in Ghana, Lets People Determine With a Text Message Whether Their Medicine is Legitimate," *Technology Review*, www.technologyreview.com/lists/innovators-under-35/2013/entrepreneur/bright-simons, accessed July 8, 2014.
363. D.-J. Jansen, R. Mosch, and C. van der Cruijsen (2013) "When Does the General Public Lose Trust in Banks?" DNB Working Paper No. 402.
364. K. Bowman and A. Rugg (2013) *Five Years After the Crash. What Americans Think About Wall Street, Banks, Business, and Free Enterprise* (Washington, DC: American Enterprise Institute).
365. VBDO (2012) "Integratie van Duurzaamheid in Toezicht Financiële Sector (Integration of Sustainability in Control on Financial Sector)," www.vbdo.nl/nl/pers/persberichten/887/integratie-van-duurzaamheid-in-toezicht-financi%EBle-sector-, accessed July 8, 2014.
366. Occupy Movement, www.occupytogether.org.
367. ASN Bank (2014) "ASN Bank in 2013: Krachtige Cijfers in Bewogen Jaar (ASN Bank in 2013: Solid Results in a Turbulent Year)," http://nieuws.asnbank.nl/asn-bank-in-2013-krachtige-cijfers-in-bewogen-jaar, accessed July 8, 2014; De Volkskrant (2013) "Triodos Bank Blijft Groeien Ondanks Crisis (Triodos Bank Keeps on Growing Despite the Crisis)," www.volkskrant.nl/vk/nl/2680/Economie/article/detail/3499787/2013/08/28/Triodos-Bank-blijft-groeien-ondanks-crisis.dhtml, accessed July 8, 2014.
368. See for instance: ABN AMRO, www.abnamro.com/nl/duurzame-ontwikkeling/particuliere-en-zakelijke-oplossingen/particuliere-oplossingen/duurzaam-sparen/index.html; ING, www.ing.nl/particulier/sparen/sparen-met-vaste-rente/groen-spaardeposito/duurzaam-sparen-bij-de-ing.aspx?first_visit=true; Rabobank, www.rabobank.nl/particulieren/producten/sparen/spaarrekening/rabogroensparen.
369. B. Brunner and B. Rowen (2007) "The Equal Pay Act," *Infoplease*, www.infoplease.com/ipa/A0763170.html www.infoplease.com/spot/equalpayact1.html, accessed July 8, 2014.
370. Lilly Ledbetter, "About," http://www.lillyledbetter.com/about.html, accessed July 8, 2014.
371. K. Pickert (2009) "Lilly Ledbetter," *Time*, January 29, 2009, http://content.time.com/time/nation/article/0,8599,1874954,00.html, accessed July 8, 2014.

372. H. Brown (2009) "Equal Payback For Lilly Ledbetter," *Forbes*, April 28, 2009, www.forbes.com/2009/04/28/equal-pay-discrimination-forbes-woman-leadership-wages.html, accessed July 8, 2014.
373. B. Brunner and B. Rowen (2007) "The Equal Pay Act," *Infoplease*, www.infoplease.com/ipa/A0763170.html www.infoplease.com/spot/equalpayact1.html, accessed July 8, 2014.
374. B. Covert (2014) "Five Years After the Lilly Ledbetter Act, How to Start Closing the Gender Wage Gap," *Think Progress*, http://thinkprogress.org/economy/2014/01/29/3223331/lilly-ledbetter-gender-wage-gap, accessed July 8, 2014.
375. Lilly Ledbetter, "About," www.lillyledbetter.com/about.html, accessed July 8, 2014.
376. Read more about SCOPEinsight on www.scopeinsight.com.
377. Ashoka, "Lucas Simons," www.ashoka.org/fellow/lucas-simons, accessed July 8, 2014.

About the author

Lucas Simons likes taking on large, complex, social challenges. His passion is sustainable agriculture. He is a change-maker and social entrepreneur, and founder and owner of two companies: NewForesight, a strategic consultancy company working on sustainable market transformation; and SCOPEinsight, a farmer organization assessment company that bridges the gap between professional farmer organizations, markets, and finance. Previously, Simons was director of UTZ Certified, a leading global certification and sustainability standard organization for sustainable and traceable coffee, cocoa, and tea.

In 2011 Simons was honored as Young Global Leader by the World Economic Forum for his commitment and accomplishments in sustainable trade, agriculture, and rural development. In 2013 he became an Ashoka Fellow in recognition of his pioneering role as social entrepreneur and systemic change-maker. For three consecutive years—2011, 2012,

2013—Simons has been ranked as one of the most influential people in sustainable development in the Netherlands.

Lucas Simons holds an MSc degree in environmental engineering from Wageningen University and an MBA degree with Merit from TIAS—School for Business and Society.

He is the proud father of three children, he is married, and lives in the Netherlands.

Index